Atomic Force Microscopy/Scanning Tunneling Microscopy 2

Atomic Force Microscopy/Scanning Tunneling Microscopy 2

Edited by

Samuel H. Cohen and
Marcia L. Lightbody
U.S. Army Soldier Systems Command
Natick Research, Development and Engineering Center
Natick, Massachusetts

Plenum Press • New York and London

Library of Congress Cataloging in Publication Data

Atomic force microscopy/scanning tunneling microscopy 2 / edited by Samuel H. Cohen and Marcia L. Lightbody.

p. cm.

"Proceedings of the Second U.S. Army Soldier Systems Command, Natick Research, Development, and Engineering Center Atomic Force Microscopy/Scanning Tunneling Microscopy (AFM/STM) Symposium, held June 7–9, 1994, in Natick, Massachusetts"—T.p. verso.

Includes bibliographical references and index.

ISBN 0-306-45596-X

1. Atomic force miscroscopy. 2. Scanning tunneling microscopy. I. Cohen, Samuel H. II. Lightbody, Marcia L. III. U.S. Army Soldier Systems Command, Natick Research, Development, and Engineering Center Atomic Force Microscopy/Scanning Tunneling Microscopy (AFM/STM) Symposium (2nd:1994: Natick, Mass.)

QH212.A78A863 1997 97-9548

502'.8'2—dc21 CIP

Proceedings of the Second U. S. Army Soldier Systems Command, Natick Research, Development and Engineering Center Atomic Force Microscopy/Scanning Tunneling Microscopy (AFM/STM) Symposium, held June 7 – 9, 1994, in Natick, Massachusetts

ISBN 0-306-45596-X

A Division of Plenum Publishing Corporation
233 Spring Street, New York, N. Y. 10013

http://www.plenum.com

10 9 8 7 6 5 4 3 2 1

Printed in the United States of America

PREFACE

This book represents the compilation of papers presented at the second Atomic Force Microscopy/Scanning Tunneling Microscopy (AFM/STM) Symposium, held June 7 to 9, 1994, in Natick, Massachusetts, at Natick Research, Development and Engineering Center, now part of U.S. Army Soldier Systems Command.

As with the 1993 symposium, the 1994 symposium provided a forum where scientists with a common interest in AFM, STM, and other probe microscopies could interact with one another, exchange ideas and explore the possibilities for future collaborations and working relationships. In addition to the scheduled talks and poster sessions, there was an equipment exhibit featuring the newest state-of-the-art AFM/STM microscopes, other probe microscopes, imaging hardware and software, as well as the latest microscope-related and sample preparation accessories. These were all very favorably received by the meeting's attendees.

Following opening remarks by Natick's Commander, Colonel Morris E. Price, Jr., and the Technical Director, Dr. Robert W. Lewis, the symposium began with the Keynote Address given by Dr. Michael F. Crommie from Boston University. The agenda was divided into four major sessions. The papers (and posters) presented at the symposium represented a broad spectrum of topics in atomic force microscopy, scanning tunneling microscopy, and other probe microscopies.

In the first session, Semiconductor Characterization and Adsorbate Characterization, topics ranged from autocorrelation analysis of AFM images to AFM and STM of chemical vapor deposited diamond-like carbon thin films to STM study of adsorbed hydrocarbons on graphite. The second session, Biological and Chemical Nanostructure, included papers on AFM techniques for imaging cells and biomolecules, real-time visualization of biopolymer degradation, enzymatic modification of a chemisorbed lipid monolayer and electron transfer reactions of organic molecules. The third session, New Developments in AFM/STM, featured papers on spectroscopic imaging with STM, advances in piezoresistive cantilevers for AFM, nanometer-scale quantitative analysis with a modified STM/field-emission source, and conductive fluctuations in tunneling junctions of a polymer film. The fourth session, AFM/STM in Materials Science, included papers on the applications of AFM in optical fiber research, nanomechanics of single crystal surfaces, and the application of magnetic force microscopy in magnetic recording.

The editors of this volume are especially appreciative of the editorial reviews of the papers by Ms. Margaret A. Auerbach and Mr. Ronald Segars. The demands of organizing and conducting this meeting were made easier by the assistance of many individuals. Ms. Mona Bray deserves a very special acknowledgment and sincerest thanks for the superior work she did in laying the groundwork for the symposium. Other duties prevented her from being directly involved, however, whenever her advice was solicited, she willingly provided us with the information we sought. Special thanks and acknowledgments go to Ms. Karen Schneider and Mr. Rhys Wyman for providing both administrative and clerical functions, and Ms.

Ethlynne Jordan for coordinating a multitude of tasks. Thanks also to Mr. Raymond Andreotti who frequently was called upon to help with solving data entry problems. Sincere thanks to Mr. David Cameron and Mr. Thomas Cook,and Ms. April Doyle, all from Defense Printing Service, for their prompt responses to all of our priority printing requests. Thanks also to Ms. Margaret Morin, Ms. Ann LaLonde, and Ms. Fran Rhymer for helping during registration and providing assistance to the attendees throughout the week. Thanks are due to Strategic Communications for the trouble-free projection of slides and viewgraphs and to Jay Weiss, who helped set up the exhibitors' displays. Special thanks to the Franklin Caterers of Framingham, MA, for providing refreshments each day and to the staff of Northeastern University's Warren Conference Center in Ashland, MA, for hosting a get-acquainted dinner for meeting attendees.

Finally, special thanks to the moderators of the daily sessions - Dr. Arnold Howard, Sandia National Laboratories, Dr. Rajan Vempati, Lamar University, Dr. C. Patrick Dunne, U.S. Army Natick RD&E Center, Dr. Jose Perez, University of North Texas, Professor Ernest C. Hammond, Jr., Morgan State University, and Dr. Timothy L. Porter, Northern Arizona University.

August 1996

Samuel H. Cohen
Marcia L. Lightbody

CONTENTS

BIOLOGICAL AND CHEMICAL NANOSTRUCTURE

Moderator: C. Patrick Dunne, U.S. Army Soldier Systems Command, Natick Research, Development and Engineering Center

NEW DEVELOPMENTS IN AFM/STM

Moderator: Jose Perez, University of North Texas

AFM/STM IN MATERIALS SCIENCE

Moderators: Ernest C. Hammond, Morgan State University
Timothy L. Porter, Northern Arizona University

KEYNOTE ADDRESS:

SCANNED PROBE MICROSCOPY

Michael F. Crommie

Boston University
Department of Physics
Boston, MA 02215

Over the last decade rapid growth has occurred in the variety of scanned probe techniques available.[1] The vast array of familiar reciprocal space probes is now joined by a multitude of real space probes. The relatively new ability to observe real space properties of systems at microscopic-length scales has found wide applicability across the physical sciences, from physics and chemistry to biology, with no end in sight. Much of this advancement has been made possible by the ready availability of commercial instruments operating in air, liquid, and vacuum. One can now even purchase high performance combined STM/AFM devices, tailored to a particular application. Despite this rapid advancement, however, there still exists a feeling that the fields of scanned probe microscopy are in their infancy, perhaps with the best yet to come. New techniques continue to be developed, and a new philosophy of experimentation has begun to take shape. Rather than just use scanned probe instruments as passive tools of surface characterization, researchers are increasingly using them to intentionally modify the systems under study. This change brings with it a great many new possibilities.

The scanning tunneling microscope began as a tool for characterizing semiconductor surfaces, and continues to be useful in this regard. The earliest investigations focused on the fundamental properties of pristine surfaces, such as the 7x7 reconstruction of Si(111),[2] the buckled bond structure of GaAs(110)[3] and the dimerized Si(001) surface.[4] These investigations took advantage of the spectroscopic capabilities of the STM to probe the real space properties of electronic levels at different energies. Later studies expanded the utility of the STM by focusing on the microscopic properties of adsorbate and metal covered semiconductors.[5] Exciting studies have also been performed of cross sectional slices of semiconductor heterostructures, including (recently) the real space mapping of delta doped layers.[6] Simultaneous advances have occurred in the study of metals, from observations of the herringbone reconstruction of Au(111)[7] to the delocalized surface state of Cu(111).[8] An exciting direction in the study of metals has been the investigations of growth phenomena for different material systems. Studies of diffusion limited aggregation,[9] surfactant mediated growth,[10] and the early stages of epitaxy[11,12] have all yielded fruitful results. Etching processes, the opposite of growth, have also been studied.[13] Novel states of matter have been investigated

Atomic Force Microscopy/Scanning Tunneling Microscopy 2
Edited by S.H. Cohen and M.L. Lightbody, Plenum Press, New York, 1997

as well, such as charge density wave systems[14,15] and the Abrikosov flux lattice in different materials[16] (most recently high-T_c superconductors[17]).

Scanned force microscopies, a slightly more sophisticated class of techniques than the STM, got a later start than the tunneling microscope. Their development and application, however, has been no less active. One advantage of the atomic force microscope over the STM is that it can observe insulating surfaces. This has allowed the study of material that could not be investigated via STM, such as diamond.[18] The ability to monitor microscopic forces now allows the measurement of friction at an atomic level.[19] Use of magnetic tips has allowed the real space measurement of magnetic force interactions and the mapping of magnetic domains and memory devices.[20] One of the most significant impacts of the development of AFM technology is that it has opened up biological systems to scanned probe investigation.[21] In situ investigations now allow the microscopic observation of living cells,[22] and one can only wonder at the variety of new explorations that will result from this.

Surface modification has always been a fact of scanned probe microscopy through inevitable tip crashes. The new excitement, though, stems from the fact that advances in technique now allow an unprecedented level of control for surface modification. This is perhaps best exemplified by Eigler and Schweizer's arrangement of single Xe atoms to spell "IBM".[23] Other researchers have shown that it is possible to remove a single Si atom from crystal surfaces,[24] modify metal reconstructions,[24] rearrange C-60 buckeyballs on a Si surface,[25] and perform various types of nanolithography.[24] On the Cu(111) surface single Fe atoms have been arranged into quantum corrals using the tip of an STM.[26] These microscopic structures scatter the 2-D surface state electrons of Cu(111), allowing real space imaging of tailor-made quantum interference patterns and the direct measurement of adsorbate scattering phase shifts.[27] Such direct control of purely quantum mechanical behavior could potentially lead to new fundamental studies of transport phenomena, adsorbate/surface interactions, localization, and magnetic behavior at the microscopic level.

One of the great hopes for the emerging topic of atomic scale construction is that new techniques will be developed for building useful microscopic devices. Atomic scale construction with scanned probe instruments represents the ultimate level of miniaturization in condensed matter systems. One might speculate that these new abilities will eventually help lead to new technologies involving ever smaller wires and switches. Perhaps tailormade molecules and even *in-situ* biological modifications are not too far down the road. For now, the field remains in its infancy, with exciting new discoveries regularly pushing back the frontier of what is possible.

1. R. Wiesendanger and H.J. Guntherodt, eds., "Scanning Tunneling Microscopy," Springer-Verlag, New York (1993).
2. G. Binnig, H. Rohrer, C. Gerber and E. Weibel, 7x7 Reconstruction on Si(111) resolved in real space, *Phys. Rev. Lett.* 50:120-123 (1983).
3. R. M. Feenstra, J. A. Stroscio, J. Tersoff and A. P. Fein, Atom-selective imaging of the GaAs(110) surface, *Phys. Rev. Lett.* 58:1192-1195 (1987).
4. R.M. Tromp, R.J. Hamers and J.E. Demuth, Si(001) dimer structure observed with scanning tunneling microscopy, *Phys. Rev. Lett.* 55:1303-1306 (1985).
5. J.A. Stroscio and W. J. Kaiser, eds., "Scanning Tunneling Microscopy," Academic Press, Inc., San Diego (1993).
6. M.B. Johnson, P.M. Koenraad, W.C. v. d. Vleuten, H.W.M. Salemink and J.H.Wolter, Be delta-doped layers in GaAs imaged with atomic resolution using scanning tunneling microscopy, *Phys. Rev. Lett.* 75: 1606 (1995).
7. C. Woll, S. Chiang, R. J. Wilson and P.H. Lippel, Determination of atom positions at stacking-fault dislocations on Au(111) by scanning tunneling microscopy, *Phys. Rev. B* 39: 7988-7991 (1988).
8. M. F. Crommie, C. P. Lutz and D. M. Eigler, Imaging standing waves in a two-dimensional electron gas, *Nature* 363: 524-527 (1993).
9. D. D. Chambliss and R. J. Wilson, Relaxed diffusion limited aggregation of Ag on Au(111) observed by scanning tunneling microscopy, *J. Vac. Sci.* B 9:2 928-932 (1991).

10. D. Kandel and E. Kaxiras, Surfactant mediated crystal growth of semiconductiors, *Phys. Rev. Lett.* 75:2742 (1995).
11. J. A. Meyer, J. Vrijmoeth and R.J. Behm, Importance of the additional step-edge barrier in determining film morphology during epitaxial growth, *Phys. Rev. B* 51:14790 (1995).
12. H. Brune, H. Roder and K. Kern, Kinetic processes in metal epitaxy studied with variable temperature STM: AG/Pt (111), *Thin solid films* 264:230 (1995).
13. E.A. Eklund, E.J. Snyder and R.S. Williams, Correlation from randomness: quantitative analysis of ion-etched graphicte surfaces using the scanning tunneling microscope, *Surface Science* 285: 157 (1993).
14. R.V. Coleman, Z. Dai and W.W. Mcnairy, Surface structure and spectroscopy of charge-density wave materials using scanning tunneling microscopy, *Applied Surface Science* 60, 485(1992).
15. R. E. Thomson, B. Burk and A. Zettl, Scanning tunneling microscopy of the charge-density-wave structure in IT-TaS_2, *Phys. Rev. B* 49, 16899-16916 (1994).
16. H. F. Hess, R. B. Robinson, R. C. Dynes, J. J. M. Valles and J. V. Waszczak, Scanning-tunneling-microscope obervation of the Abrikosov flux lattice and the density of states near and inside a fluxoid, *Phys. Rev. Lett.* 62, 214 (1989).
17. I. Maggio-Aprile, C. Renner, A. Erb, E. Walker and O. Fischer, Direct vortex lattice imaging and tunneling spectroscopy, *Phys. Rev. Lett.* 75:2754-2757 (1995).
18. G.J. Germann, S.R. Cohen and G. Neubauer, Atomic scale friction of a diamon tip on diamond (100) and (111) surfaces, *J. Appl. Phys.* 73:163-167 (1993).
19. M.D. Perry and J.A. Harrison, Universal aspects of the atomic scale friction of diamond surfaces, *J. Phys. Chem.* 99:960-65 (1995).
20. D. Sarid, "Scanning Force Microscopy" Oxford University Press, New York (1991).
21. P. K. Hansma, V. B. Elings, O. Marti and C. E. Bracker, Scanning tunneling microscopy and atomic force microscopy: Application to biology and technology, *Science* 242:209-216 (1988).
22. J. H. Hoh and P. K. Hansma, Atomic force microscopy for high resolution imaging in cell biology, *Trends in Cell Biology* 2: 208-213 (1992).
23. D. M. Eigler and E. K. Schweizer, Positioning single atoms with a scanning tunneling microscope, *Nature* 344: 524-526 (1990).
24. P. Avouris, ed., "Atomic and Nanometer-Scale Modification of Materials: Fundamentals and Applications" Kluwer Academic Publishers, Boston (1993).
25. P.H. Beton, A. Dunn and P. Moriarty, Manipulation of C_{60} molecules on a Si surface, *Appl. Phys. Lett.* (in press).
26. M.F. Crommie, C. P. Lutz and D. M. Eigler, Confinement of electrons to quantum corrals on a metal surface, *Science* 262: 218-220 (1993).
27. E.J. Heller, M.F. Crommie, C.P. Lutz and D.M. Eigler, Scattering and adsorption of surface electron P. Avouris waves in quantum corrals, *Nature* 369:464-466 (1994).

SEMICONDUCTOR CHARACTERIZATION AND ADSORBATE CHARACTERIZATION

Moderators: Arnold J. Howard, Sandia National Laboratories
Rajan K. Vempati, Lamar University, Texas

SCANNING TUNNELING MICROSCOPY FOR VERY LARGE-SCALE INTEGRATION (VLSI) INSPECTION

Shane Y. Hong

Department of Mechanical Engineering
Columbia University
New York City, NY 10027

Abstract: This paper discusses the efforts toward overcoming the difficulty in the inspection and measurement of submicrometer structures of very large-scale integration (VLSI) circuitry. The microelectronic industry has been using finer line-width and spacing to build denser circuitry. When the lines and spacings are smaller than the light wavelength, the traditional optical microscope is no longer able to image the fine lines due to light diffraction.

A new method of surface profiling and dimensional measurement of submicron VLSI structures using scanning tunneling microscopy (STM) has been successfully tested. An improved technique for etching sharp and slender STM probes has been developed, enabling the STM to be applied to high-density, high-rise microelectronic structures and reducing the measurement error caused by the probe geometry. Probes with an ideal tip geometry, with a tip angle less than 3° and a radius of 0.03 μm within 1 μm from the tip, can be consistently produced and are believed to be state-of-the-art. Furthermore, the methods of side wall profiling and true profile reconstruction are developed to avoid the probe geometrical effect, making it possible for STM to obtain an accurate topographical profile without cleaving the sample and viewing it from the edge as needed by scanning electron microscopy (SEM).

Compared to scanning electron microscopy SEM, STM operates in air, provides three-dimensional imaging and yields better resolution; therefore, STM has a greater potential than SEM. The measurement error caused by the geometry of the probe, the only outstanding issue affecting STM accuracy, is explored in detail. The same techniques developed in this study can also be applied to atomic force microscope (AFM), a derivative of STM, for profiling and measuring nonconductive samples.

INTRODUCTION

This paper discusses the efforts toward overcoming the difficulty in the inspection and measurement of sub-μm structures of VLSI circuitry. The microelectronic industry has been using finer line-width and spacing to build denser circuitry. When the lines and spacings are

Atomic Force Microscopy/Scanning Tunneling Microscopy 2
Edited by S.H. Cohen and M.L. Lightbody, Plenum Press, New York, 1997

smaller than the light wavelength, the traditional optical microscope is no longer able to image the fine lines due to light diffraction. Currently, scanning electron microscope is used for mapping and measuring microelectronic line/spacing structure of 0.25 μm or below. However, SEM has its own shortcomings and limitations.[1] The purpose of this study is to find an alternative tool for topographic profiling and dimensional measurement of submicrometer structures of microelectronic circuitry.

With the invention of the STM in the early 1980s,[2-4] considerable progress has been made in improving it as an increasingly used tool.[5] Its high resolution is an attractive reason for us to conduct a study to determine if it can be applied to the VLSI inspection. STMs have been widely used to acquire atomic structures such as graphite, silicone, metal and biological structures such as DNA. However, there are problems in applying STM in the topographical profiling and line-width measurement of VLSI structures: (1) different from previous STM applications, which require atomic resolution in the order of 10^{-10}m, VLSI structures are in the order of 10^{-6}m, which require a large scanning range for STM; (2) the VLSI structures are composed of sharply rising step structures (lines) and deep, narrow grooves (spacings) with vertical side walls; (3) the VLSI structures include conductors, insulators and semiconductors; STM can only operate on conductive or semi-conductive surfaces, and (4) the probe used by STM is critical to the accuracy of measurement.

To do this, the feasibility of using STM to inspect microelectronic structures must first be demonstrated, then its capability and limitations have to be investigated. Consequently, its potential can be evaluated by comparing it to current SEM. If STM is valuable for VLSI applications, we should find a way to resolve its shortcomings.

This paper discusses the issues and reports the successes of applying STM to the inspection of VLSI structures. It covers the experiments and the results of testing parallel line structure, test pattern and multilevel VLSI circuitry. It also discusses the measurement errors caused by the scanning probe geometry and reveals the remedy. An improved etching method for controlling the ideal characteristics of the STM probe is introduced. The technique for sidewall profiling and for reconstructing the true profile is described. A method of probe self-calibration is explored for measurement error compensation. Finally STM is evaluated by comparing it with SEM.

FEASIBILITY STUDY

Test Sample and the Machine

The sample was composed of a wafer patterned with groups of parallel photoresist lines and electrical line-width tester patterns which represented important features of VLSI circuitry. Two types of samples were used: one with 0.75 μm photoresist and one with 0.40 μm aluminum patterns. Since the STM could not accept an entire wafer, the samples were sliced to 5 mm x 8 mm from 4" wafers. They were coated with a conductive layer (100 Å of gold). There was no specific reason for coating with gold, but it was a convenient way to form a conductive layer in the laboratory.

An etched tungsten probe was used in the Nanoscope II STM to scan over the sample. The scan size (range) depended mainly on the head and voltage range chosen. The picture obtained by STM scanning was composed of 200 points per line x 200 lines per frame and for

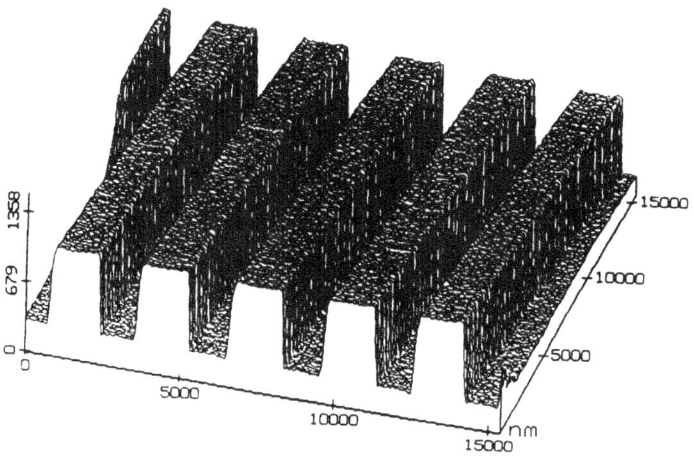

Figure 1. STM plot of 1.5 parallel lines as 0.75 μ thick.

higher resolution 400 x 400. The time needed to construct a picture was therefore the scanning rate times the number of lines per frame. For most testing, the scanning rate was set to 1 Hz; it took 3.3 minutes to get a picture. For line-width measurement, which does not need to have a whole picture taken, it could be done in a matter of seconds.

RESULTS

Test Result: Parallel Lines

Figure 1 shows the three-dimensional line plot of the STM results of ~1.5 μm parallel lines of ~0.75 μm thickness. This figure also shows how well an STM can trace the vertical wall.

The scan result of the STM is indicated in Figure 2 for ~1 μm parallel lines in the large scanning mode, and Figure 3 is a more detailed picture. For ~0.75 μm parallel lines, the scan result of the STM is shown in a three-dimensional shaded surface image using solid modeling techniques given in Figure 4. This picture was taken from the color display terminal of the Nanoscope II. The image of the number ~(0.75) on the sample indicating the line-width was also obtained by STM scanning.

For the narrowest lines, ~0.5 μm wide and ~0.5 μm apart, the STM image (Figure 5) is shown in a 3-D line plot with shading indicating height level. The irregularity of the line edge was real, caused by a processing problem of the sample, which was verified by the blurred, irregular lines imaged from an SEM with a magnification of 10,000 (see Figure 6).

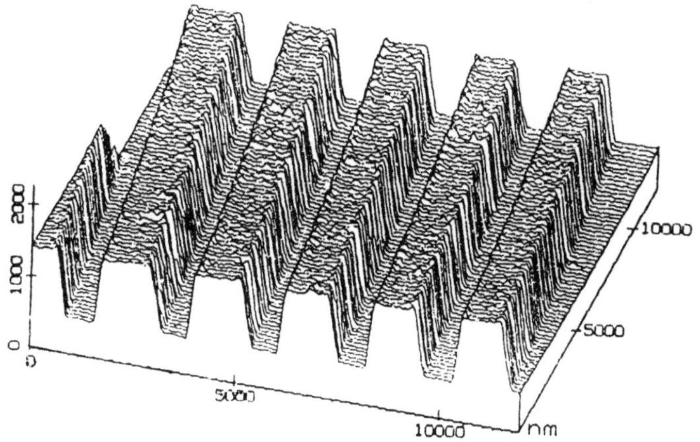

Figure 2. STM plot of 1 μm parallel lines in large scan mode.

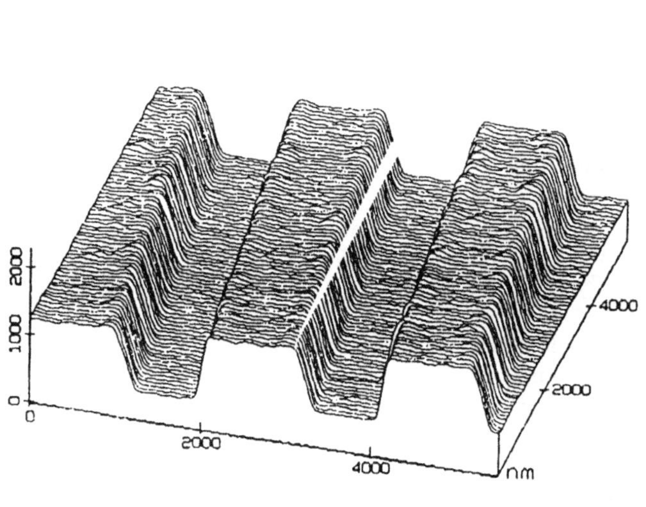

Figure 3. STM detail scan of 1 μm parallel line.

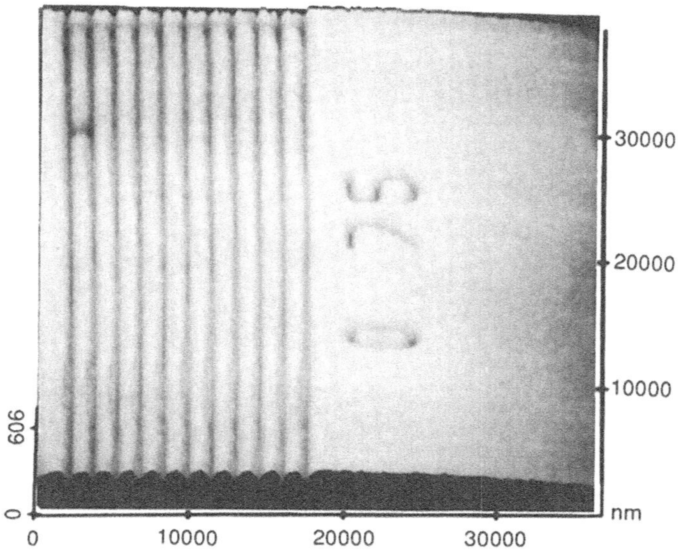

Figure 4. 3-D shaded surface STM image of 0.75 μm parallel line.

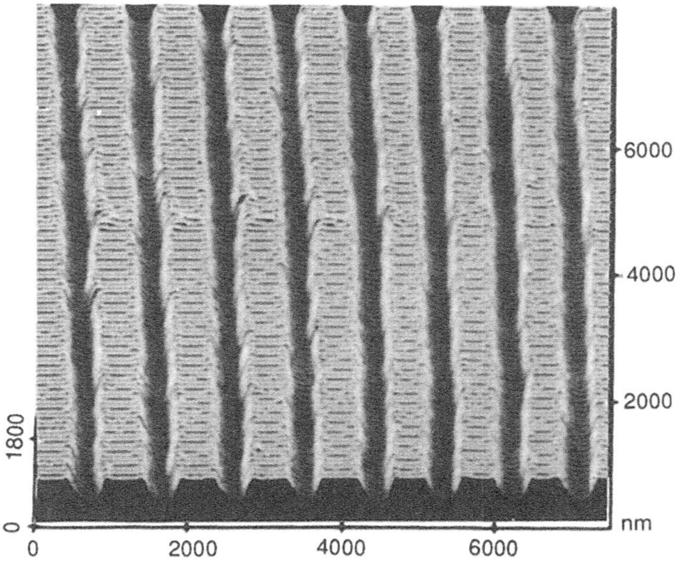

Figure 5. Parallel 0.5 μm lines test the limit of STM capability and the quality of sample preparation.

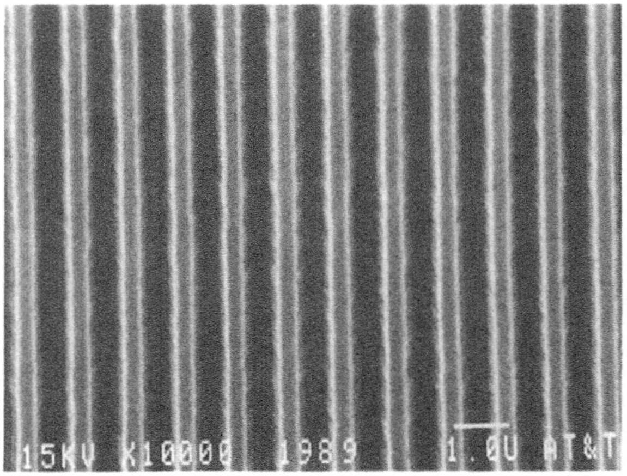

Figure 6. SEM image confirms the irregularity of the 0.5 μm lines sample.

The Nanoscope II provided the capability of image sectioning. A STM image of a single line of 0.5 μm, with its label, is shown in Figure 7. The sample is patterned by aluminum, which is about 0.375 μm thick. The lower left picture is a top view, with height shown by difference in color intensity. The plane of section is chosen by indicating two points on the top view and the cross section profile is created, shown on the upper plot. Distance, either vertical or horizontal, can be measured by setting the two cursors in either plot.

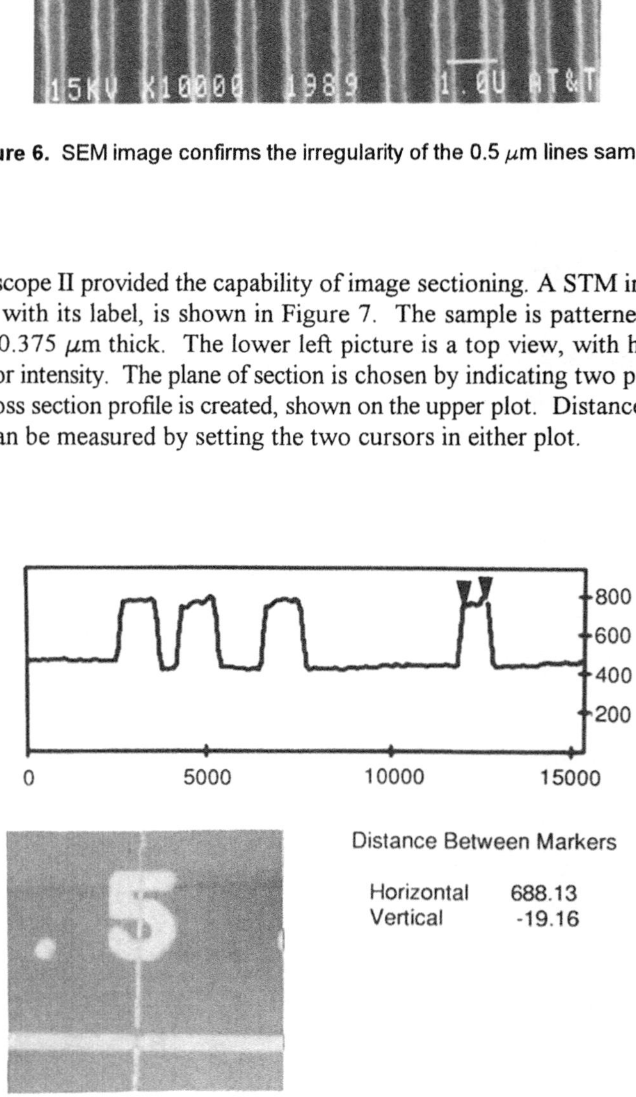

Figure 7. STM image of a single 0.5 μm line with the label.

STM seems to have the capability of acquiring a good image of smaller features. However, the 0.25 μm lines and spaces on the sample were not well developed; this was verified by a SEM. Therefore, no STM result was associated with a 0.25 μm line. The issues involving smaller dimensions are waiting further study.

Test Result: Circuit Patterns

For researchers to see a recognizable circuit pattern requires that the image cover a large area. The STM scan images, described below, are related to the electrical testing pattern. This testing has been designed for measuring proximity effects in electron beam lithography.

Figure 8 (a) shows the photoresist pattern of an area indicated in Figure 8 (b), which includes a 5 μm bent line from a corner of a square testing pad, an elbow, and half of the letter "o".

Figure 9 shows a series of scans over the electrical probe testing pattern. The corresponding parts of the features in Figure 9 (a), (b), (c) are indicated in Figure 9 (d). It also shows the reproducibility and the repeatability of the STM.

Test Result: Multilevel VLSI Structure of RAM Memory Device

A finished VLSI product is usually constructed by a multilevel structure with different layers of conductor, insulator and semiconductor. Figure 10 shows the STM images of a small spot in the circuitry of 1M-bit dynamic random access memory. These three-dimensional, shaded surface images of STM reveal a resolution superior to SEM. Some small faults in the circuitry can be easily identified.

SUMMARY FROM TEST RESULTS

As the test results show, the STM can be used for VLSI metrology and demonstrates superior features over other commercial metrology tools. The top surface of the feature can be revealed accurately; therefore, the measurement of the line-width at the top surface should be better than that obtained using the optical method or SEM. Also, the best height measurement obtained is, without a doubt, from STM, which requires no cleaving or side-viewing.

PROBE GEOMETRY EFFECT

In measurement systems that use probes, such as a coordinate measuring machine, the geometry of the probe is critical to the accuracy of measurement. In the case of STM, the probe can not be uniformly fabricated because the tip is so small. However, it is not dimensionless, and the scanning result by STM tends to get a wider line-width for the raised line, while the width of the trench tends to be narrower. When the probe is well fabricated, its tip can be approximated either by a hemisphere with radius r and a cone with angle α, or by a second order equation. The profiles obtained from STM scanning deviate from the true profiles

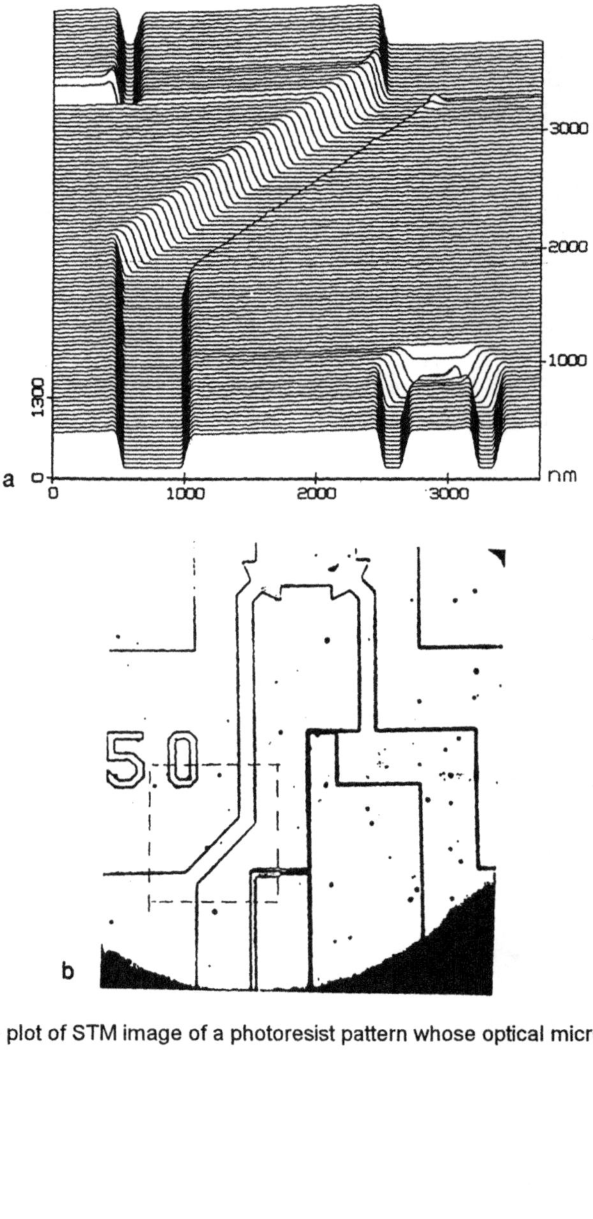

Figure 8. (a) Line plot of STM image of a photoresist pattern whose optical microscope image is shown in (b).

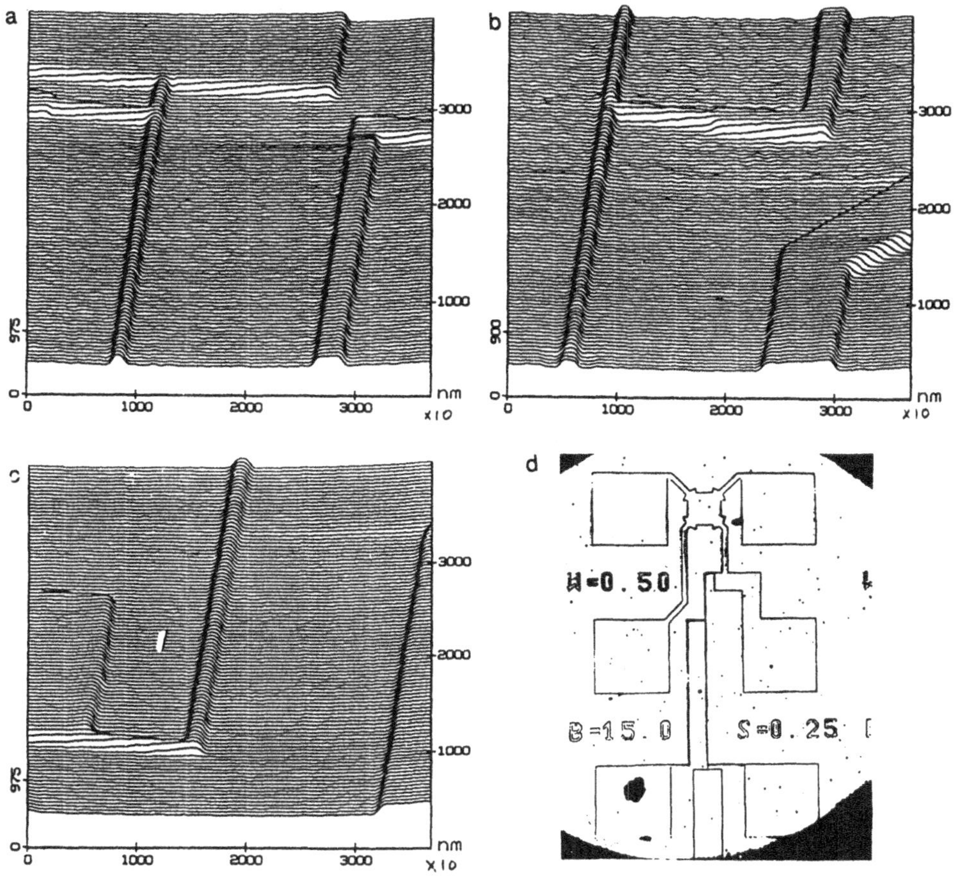

Figure 9. The sequential STM images scan over an area shown in (a), (b), (c),compared with the optical microscope image shown in (d).

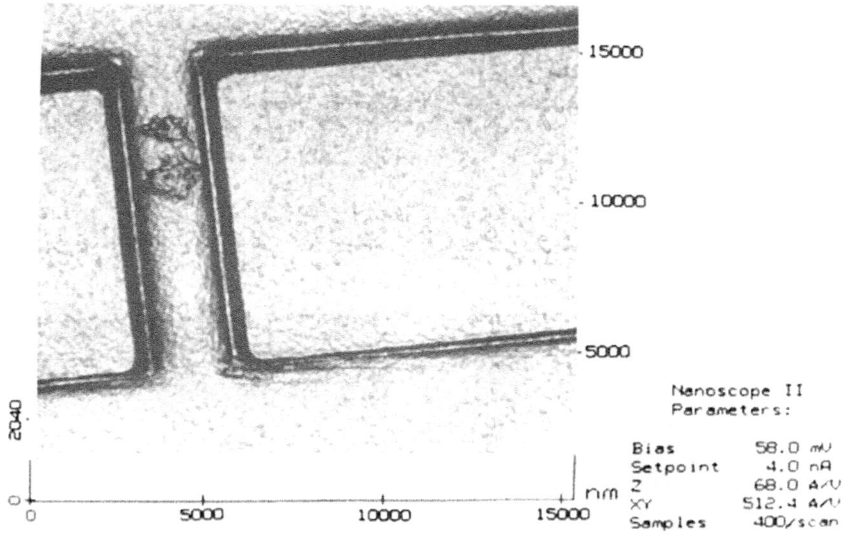

Figure 10. STM image of a spot at 1Mbit DRAM circuitry. Data taken 1989; buffer 1, rotated 0°; XY axes (nm), z axis (nm).

of VLSI structure pattern. It generates a tapered side wall profile for perfectly vertical walls or undercut side walls (in which the bottom width of the raised structure is smaller than the top width). Therefore, there is an error, caused mainly by the half cone angle of the probe and, in a minor way, caused by the hemisphere radius of the probe tip.

The probe geometry will cause more profile distortion when STM is used to scan across the deep and narrow trenches. If the trenches have height h and width w, and a probe has tip radius r and probe angle α, the probe tip will not touch the trench bottom when:

$$w \leq 2[h + r(\frac{1}{\sin\alpha} - 1] \tan\alpha \tag{1}$$

Since the scanning probe is not dimensionless, the STM inherently has measurement inaccuracy due to the tip geometry. In order to improve the measurement accuracy and to enable the STM to inspect the finer trenches or holes (especially those below 0.25 μm), the following efforts are necessary: (1) improving the probe tip fabrication technology to produce probes with an ideal, well-controlled geometry (2) developing the scanning and reconstruction method to reveal the true profile of the side walls (3) inspecting and calibrating the probe by STM itself, and (4) providing the dimensional and geometrical error compensation. The following sections address these four tasks.

REFINEMENT OF PROBE-SHAPING PROCESS

As the integrated circuit technology progresses toward miniaturization, the industry has worked toward fitting as many circuits as possible onto one wafer. Therefore, smaller line-widths are being produced. In order for the STM probe to reach the corners of the grooves and make accurate measurements, the probe should ideally have a small angle as well as an extremely sharp tip (i.e., small tip radius). The height h of the microelectronic structure is normally 0.7 μm for the photoresist layer and about 0.375 μm for the conductive layer. Since the VLSI industry is now developing 0.25 μm technology, the width w of the trench or the diameter of the via hole is targeted for less than 0.25 μm. In order for the probe to touch the bottom, by using Eq. (1), the desirable probe geometry should have an angle less than 3° and a radius less than 0.05 μm, within 1 μm of the tip.

Commercially, STM probes are made of Pt-Ir or tungsten, and are either mechanically formed (shear at an angle) or electrochemically etched in sodium nitrite or potassium hydroxide. A mechanically formed probe does not have a uniform shape and is not useful for VLSI inspection. The most common method used has been the chemical etching by inserting a small portion (2-5 mm) of a tungsten wire (10 mil diameter) vertically into a solution (5%) of potassium hydroxide (KOH), and applying an AC or DC voltage (about 30V) across the wire and solution. Commercial STM probes have a wide scattering of the geometry; few can meet the requirements for VLSI inspection. This wide scattering of the radius and angle distribution is not ideal for VLSI application. Although the technique used to make these probes is relatively simple, it has been difficult to consistently produce probes with the desired characteristics. It is necessary to develop a controlled, electrochemical etching process to produce an ideal probe geometry.

Methods to produce a sharp tip for Scanning Tunneling Microscopy (STM) or Field Ion Microscopy were suggested in two previous efforts by P.J. Bryant et al.[8] and H. Morikawa, et al.[9]

(1) The technique by P.J. Bryant et al.[8] to produce sharp, work-hardened asperities included the following steps: (a) A tungsten wire electrode was tapered near the lower end by direct electrochemical polishing in a 3%-5% KOH solution. A 20V ac potential was applied for initial shaping. (b) The tip region was then protected from etching by inserting it into an air column inside a glass capillary of 1.3 μm i.d. which was trapped due to the high surface tension of the KOH solution. (c) Etching of the shank proceeded, until the tensile strength of the notched region could no longer sustain the weight of the small lower end. The applied potential was reduced toward zero as the tip shank diminished in size. (d) Plastic deformation and work hardening during tensile shearing of the shank should produce small asperities. The smallest tip diameters produced were reported as approximately 50 Å and were obtained from transmission electron micrographs. This technique produced a sharp tip and was good for the atomic structure study of STM, but it did not generate the slender shape and small cone angle desired for the VLSI application.

(2) H. Morikawa and K. Goto[9] attempted to improve the AC method by controlling the applied voltage, frequency, wave shape, and wave count. They found that the probe shape depended on the frequency of the alternative current and that low frequency was remarkably effective in reducing the radius of the wire in the case of the polishing Mo with 5% KOH aqueous solution at 12 V ac. The exact geometry of their result was not reported. Our study found that their conclusion was misleading.

Since probe geometry is the important issue in applying STM to VLSI inspection, the author and his assistant[10] have done a study in order to understand the probe shaping mechanism and to find the parameters for precise control. The bubbling effect and the field effect in the etching process were assumed as two possible theories to explain the shape forming mechanism. Experiments were needed to verify the assumption. Once the probe shaping mechanism was identified, we could then control the etching parameters to form the desirable geometry for STM probes.

The key factor in the bubbling theory is the distribution of the gas bubbles around the tungsten wire during etching. The smallest amount of bubbles occurs at the lower end of the tungsten wire. The bubbles propagate upward along the sides of the wire. Thus the density of the bubbles is the highest at the sides toward the top of the wire. The bubbles act as a shield for the wire and reduce the contact between the KOH solution and tungsten; therefore, the etching reaction is inhibited. The area having the smallest quantities of bubbles, the lower end, etches the most rapidly, while the sides etch at a slower rate.

The field effect theory is based on the argument that a stronger electrical field occurs at the end of a wire because of the larger local curvature. The reaction between the wire and KOH solution is faster at the end than at the sides of the wire.

Several experiments were designed to test these theories. The results of the studies indicated that the bubbling effect greatly influenced the shape of the etched tips. The field effect was less significant to the geometry of the tip, unless the electrode in the KOH solution was located close to the tungsten wire being etched. However, it was very difficult to accurately control the potential field to achieve the desirable probe geometry.

Since etching a slender and sharp tip was our goal, we modified the electrochemical etching process to fully utilize the bubbling effect theory. By lowering the voltage applied to the tungsten wire and the solution, a better distribution of the bubble density was achieved.

Our experiment of applying a high voltage (greater than 10V) resulted in a sharp tip and a larger angle compared to a small voltage (around 5V) which yielded a more slender probe with smaller angles. By using 5 volts ac, submerging 4 mm of the wire (0.010 inch diameter), and etching 1.5-2 mm of tungsten from the end, we were able to consistently, though still not identically, etch sharp and slender tips with a hemisphere radius between 0.02 and 0.05 μm and angles between 2.5° and 5°. The tip may have been sharper than these readings, but could not be verified due to the limitation of SEM resolution and focusing capability.

SELF-CALIBRATION OF PROBE GEOMETRY

As indicated before, the accuracy of the measurement of STM greatly depends on the probe geometry. A sharp and slender probe will help reduce the measurement error. Although the control of the probe geometry has been improved as described in the previous section, it is still impossible to produce identical probes in nano scale. The method to compensate the error caused by the probe geometry needs to be developed. In order to provide error compensation, the geometry of a specific probe must be known. To avoid using an inaccurate instrument to calibrate a system for more accurate measurements, obtaining the probe geometry from SEM is not a good choice, because SEM has limited resolution and focusing capabilities. Since STM is more precise than SEM, it is ideal to use STM to calibrate the probe geometry.

The geometry of a STM probe tip can be approximated from the radius of the tip, r, and the taper angle, α. Using the probe to be calibrated, STM scans over a "master" sample in which the dimension and geometry is accurately known. The master sample includes two major features: one groove with vertical walls and another groove with tapered walls. The side wall of the latter groove is tapered with known angle θ, in which θ must be greater than probe angle α. A good example is: $\theta = 45°$.

The STM scanned result will show that the width of the bottom flat is w' instead of the true width w. The tip radius of the probe can be obtained by:

$$r = \frac{1}{2}(w - w')\cot\frac{\theta}{2} \tag{2}$$

Then the STM, using the same probe, scans across the groove with the vertical walls which have a wall height of h and a width of w_1. If the measured width of the groove bottom is w_1', and if r is the tip obtained from Eq. (2), the following relationship can be found:

$$w_1 - w_1' = 2[h - r + \frac{r}{\sin\alpha}] \tan\alpha \tag{3}$$

The solution for this equation is:

$$\alpha = -\sin^{-1} \frac{r}{\sqrt{(\frac{w_1 - w_1'}{2})^2 + (h-r)^2}} + \tan^{-1} \frac{h-r}{(\frac{w_1 - w_1'}{2})} \tag{4}$$

Through these procedures, the geometry of the specific probe, both r and α, can be determined. During the probe calibration process, if the contour obtained from scanning the master sample is irregular or deviating from the expected contour too much, then the probe is poorly made and should be discarded.

SIDEWALL PROFILING

During the manufacturing process of microelectronic circuits, the line-width must be checked regularly, and sometimes the structure profile and the photoresist profile have to be checked in process for quality control. In order to get the profile of the VLSI structure, SEM will need the sample cleaved and mounted sideways to be scanned. This process is destructive and also requires good skills. The scanning result by STM is a three-dimensional topography of the scanned surface, yielding more information than SEM can provide. As shown in Figure 1 to Figure 9, it is quite easy to get a profile of a microelectronic structure which can be done either (1) by disabling one of the x or y axis scans parallel to the line structure, or (2) by sectioning of the three-dimensional image using available computer software.

However, due to the probe geometry effect, the profile resulting from STM scanning may deviate from the true profile, especially for the vertical wall and the under-cut wall. This profiling error can be reduced if the probe is sharp and slender. Unfortunately, it is impossible to create a probe with no radius and an angle at the tip; thus, the error always exists but should be minimized. In addition, as in the case of the undercut wall of an inverse trapezoid structure, the true profile cannot be revealed unless a method of sidewall profiling is developed.

To overcome the tip angle obstruction and to measure the side wall, the side wall can be scanned by (1) tilting the sample at an angle slightly greater than the tip angle (2) tilting the scanning head in a similar manner, or (3) simply bending the probe at a similar angle. The first two approaches are preferred.

This approach is useful only when the trench (groove) is large enough to allow the probe to reach the trench button, that is:

$$w > r(1 + \frac{1}{\cos(2\alpha+\varepsilon)}) + (h-r)\tan(2\alpha+\varepsilon)$$

$$= r[1 + \frac{1}{\cos(2\alpha+\varepsilon)} - \tan(2\alpha+\varepsilon)] + h\tan(2\alpha+\varepsilon) \qquad (5)$$

where

- r = the tip radius
- 2α = the cone angle, twice the probe angle
- ϵ = the clearance angle between the sidewall and the probe
- h = the wall height, either the thickness of the photoresist or the circuit layer.

Even if the trench is wide enough, the scanning result by tilting the probe or the sample will improve one side wall profiling but will worsen the other. Therefore, it requires two scanning passes by tilting from different directions.

TRUE PROFILE RESTORATION AND DIMENSIONAL ERROR COMPENSATION

STM is very accurate for measuring the top surface and the depth. Using the top surface and the depth as references, the profiles of the three passes with different tilting directions can be superimposed to restore the true side profile as indicated in Figure 11. However, this profile still has dimensional errors for which compensation must be made.

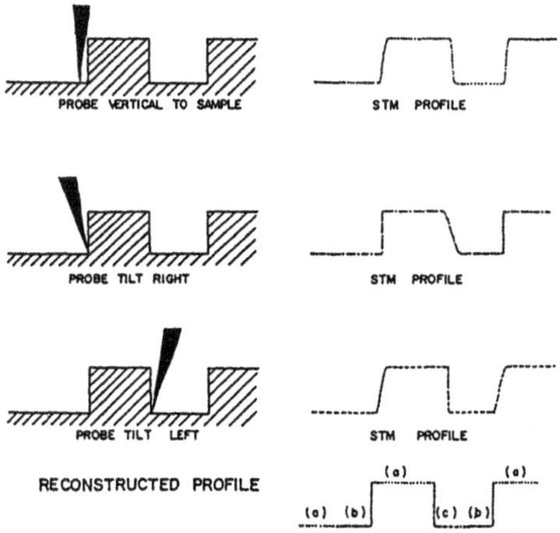

Figure 11. Illustration of the restoration method by superpositioning three profiling.

After knowing the geometry of the probe being used by self-calibration, the measurement results for the new sample inspected by STM can be adjusted mathematically. If the incline angle of the sidewall of the structure is larger than the tip angle α, the profile θ obtained from the scanning reflects the true angle. The dimension of the groove bottom can be inversely calculated by using Eq. (2):

$$w = w' + 2r \cot\frac{\theta}{2} \tag{6}$$

However, if the slope of the sidewall exactly coincides with the probe angle, the sidewall profiling and reconstruction of the true profile must be done before doing the error compensation.

COMPARISON BETWEEN SEM AND STM

Scanning Electron Microscopy has long been regarded as the most precise tool for VLSI metrology. This study, however, indicates that STM has the potential for replacing SEM. From the compiling of commercial data, publication and personal experience, a comparison between SEM and STM is listed as follows[7] :

Resolution: SEM has a normal resolution of 2-20 nanometers, although an ultra high voltage electron microscope such as JEOL ARM 1000 has a 0.16 nm resolution at 1 million volts[9]. On the other hand, STM has been used mainly for atomic resolution in the sub-Angstrom region.

Picture: The SEM picture, by nature, is two-dimensional; but the STM image is three-dimensional, which yields more information and makes Z measurements possible without cleaving.

Calibration: The SEM had no line-width standard until recently when the pitch calibration method was used. The STM uses a highly ordered pyrolytic graphite (HOPG) sample for atomic calibration and optical grating for large-scan calibration.

Cost: The SEM metrology tool costs approximately $850,000 for a Hitachi 7000, while a commercial STM such as the Nanoscope II, made by Digital Instrument, costs $75,000. These dollar amounts tend to be misleading, because they are not really comparable. Considerable modification and automation efforts must be done to the STM to be a VLSI metrology tool. Because it has no vacuum system, dedicated electron lenses or high voltage sources, the STM seems to be much cheaper to construct.

Sample Requirement: There is a charge-up phenomenon when the SEM is used on a nonconductor. A conductive coating on the wafer is beneficial but destructive. Low-voltage SEM can reduce the electron charge and will reduce the resolution[1]. The sample for STM must be a semiconductor or conductor with a resistance smaller than 1 Mega Ohm. Coating is usually required.

Operation Environment: The SEM operates in a vacuum and is sensitive to vibration and electromagnetic fields. STM can operate in air, liquid or vacuum; it is not affected by the electromagnetic fields. However, it is susceptible to vibration as is the SEM.

Measurement Error: Since the SEM is more complex, more sources can cause measurement errors: scan linearity, magnification compensation, lens hysteresis, accelerating voltage, operator factor, instrument maintenance, sample charging, detector type and location, and sample contamination effect. The measurement error sources for STM include probe geometry, piezo linearity and hysteresis.

It is clear that STM, in many respects, is superior to SEM. STM can at least supplement the weakness of SEM, if it cannot fully replace SEM.

CONCLUSIONS

With the goal of finding alternative measurement instruments for sub-μm VLSI structures, the author has studied the scanning tunneling microscope. The following conclusions are drawn:

1. Surface profiling and dimensional measurement of sub-μm VLSI structures by STM was successfully tested.
2. Compared to SEM, STM operates in air, provides three-dimensional imaging and yields better resolution, therefore, STM has greater potential than SEM.
3. The only outstanding issue affecting STM accuracy has been identified as the measurement error caused by the geometry of the probe.
4. An improved technique for etching sharp and slender STM probes has been developed. Tip radius of 0.03 µm and a tip angle less than 3° can be consistently produced, believed to be state-of-the-art.
5. A method of side wall profiling and reconstruction has been developed to avoid the probe geometrical effect. It enables STM to obtain accurate topographical profiles without the need to cleave the sample and view it from the edge, as SEM requires.
6. The probe geometry can be self-calibrated by scanning over a master sample. The STM measurement error due to the probe geometry can therefore be compensated by using a mathematical model.
7. The same techniques developed in this study can also be applied to AFM, a derivative of STM, for profiling and measuring nonconductive samples.

ACKNOWLEDGMENTS

The author thanks AT&T Bell Laboratories, Murray Hill, NJ where this study was done. Many colleagues' help, such as test samples provided by Elaine Kung, SEM pictures taken by Alexander Timko, RAM samples supplied by AT&T Microelectronics-Orlando, and the probe etching study assisted by Mae-Mae Shieh, are acknowledged. Digital Instruments, Santa Barbara, CA gave technical support when the author ran the experiment using Nanoscope II.

REFERENCES

1. M. T Postek and D.C. Joy, Microelectronics dimensional metrology in the scanning electron microscope, *AT&T Bell Lab. Technical Memorandum*, TM #11521-860804-37.
2. G. Binnig, H. Rohrer, C. Gerber, and E. Weibel, Tunneling through a controllable vacuum gap, *Appl. Phys. Lett.* 40:178-180 (1982).
3. G. Binnig, H. Rohrer, C. Gerber, and E. Weibel, Surface study by scanning tunneling microscopy, *Phys. Rev. Lett.* 49:57-60 (1982).
4. G. Binnig, H. Rohrer, C. Gerber, and E. Weibel, Vacuum tunneling, *Physics* 109 and 110b:2075-2077 (1982).
5. Y. Kuk and P.J. Silverman, Review: Scanning tunneling microscope instrumentation, *Rev. Sci. Instrum.* 60:165-180 (1989).
6. P.K. Hansma and J. Tersof, Scanning tunneling microscopy, *J. Appl. Phys.* 62:R1-R23 (1987).
7. S.Y. Hong, Scanning tunneling microscope application in VLSI submicron metrology," *Bell Lab. Technical Memorandum*, TM #52126-890817-21, 1989.
8. P.J. Bryant, H.S. Kim, Y.C. Zheng, and R. Yang, Technique for shaping scanning tunneling microscope tips, *Rev. Sci. Instrum.* 58:1115 (1987).
9. Hiroski Morikawa and Keisuke Goto, Reproducible sharp-pointed tip preparation for field ion microscopy by controlled ac polishing, *Rev. Sci. Instrum.* 59: 2195-2197 (1988).
10. Mae-Mae Y. Shieh, Summer project report for tungsten probe etching, AT&T Bell Laboratories, *Internal Memorandum for Record*, Aug. 11, 1989.

SCANNING TUNNELING MICROSCOPY-BASED FABRICATION OF NANOMETER SCALE STRUCTURES

Munir H. Nayfeh

Department of Physics
University of Illinois at Urbana-Champaign
Urbana, Illinois 61801

Abstract: We describe several STM-based techniques that we have developed for the fabrication of nanometer scale structures at room temperature. The techniques utilize processing with the biasing voltage/current of the tip of the microscope. Tunable laser radiation coupled to the gap induces multiphoton excitation or ionization processes of precursor gasses thus providing material selectivity to the process. We have made structures whose sizes range from a few hundred nanometers down to the size of individual atoms or molecules, on graphite, chemically passivated silicon, photoresist coated silicon, and organometallic-coated silicon surfaces. On the other hand, at small enough tunneling gaps, the chemical potential collapses allowing the tip to suck material off the surface, hence producing grooves. We have been able to fabricate continuous micrometer long lines of smallest widths ever (as small as 40 Å). We used this capability to fabricate all sorts of two-dimensional patterns: triangular, rectangular, circular, parallel lines, grids, and others in the shape of alphabets. In addition, we are in the process of integrating this capability with novel molecular beam epitaxy (MBE) methods to fabricate and analyze two and three dimensional nanometer scale structures such as quantum wires and dots, quantum gratings, arrays of quantum dots etc. We are presently using these techniques to construct and test the quantum interference transistor, a micrometer size metal oxide semiconductor field effect transistor (MOSFET) with a nanometer scale grating or grid embedded in its gate area. These advances have important implications to mass storage of information, which may lead to great reductions in the sizes of electronic circuits and devices.

INTRODUCTION

Recently the characteristics of fabrication with electron or ion beams, have been refined by using the corresponding beams of the tunneling gap of a scanning tunneling microscope (STM), a device that was recently invented for observation of surfaces with atomic resolution.[1,2] For example, there have been experiments that used the scanning tunneling microscope (STM) to make nanometer scale features on chemically prepared surfaces.[3-5] In these experiments voltage pulses imposed across the tunneling gap were used to generate

Atomic Force Microscopy/Scanning Tunneling Microscopy 2
Edited by S.H. Cohen and M.L. Lightbody, Plenum Press, New York, 1997

features. It was found that the process proceeds from ambient conditions to vacuum conditions. In the vacuum study[3,5] it is believed that raised features on the surface are produced by chemical reactions in the contamination resist, however, quantitative characterization of the process is not available. On the other hand, the air study[5] indicates that the features are a result of differences in the electronic properties between the modified and unmodified regions brought about by oxidation rather than real topographic structures. Other experiments have utilized transplanting atoms or clusters of atoms from the tip to the surface[6] or from the surface to the tip and back to the surface.[7] These procedures succeeded in generating structures of nanometer scale dimensions.

On the other hand, direct production of patterned features on a substrate using localized laser initiated chemical reactions has long being employed.[8] In direct writing, an ultraviolet or visible laser beam is focused onto the surface of a semiconductor wafer placed in the appropriate reagent gases. Atoms that may either etch, deposit on, or dope a solid surface are produced. If the laser wavelength is in the visible region and the substrate absorbs light at the laser wavelength, surface heating occurs and the reactions are initiated by thermal chemistry. The writing resolution is limited by the size of the laser spot to 0.5 micrometer diameter; however it is not possible to get spots smaller than this because of limits imposed by diffraction. Because of their shorter wavelengths, ultraviolet light can be focused to submicrometer dimensions before the limits imposed by diffraction are reached, so photolithography with ultraviolet rather than visible light offers a decrease in the size of features that can be produced by the lithographic process. By exploiting some nonlinear properties of the interaction of radiation with matter, Ehrlich and Tsao[8] wrote doped lines 0.2 micrometer wide in a silicon substrate, thus pushing the technique to its utmost limit. While conventional techniques are satisfactory for most production operations, there are cases where it would be convenient to produce the patterned layer directly. These cases include the repair, design, and modification of circuits. Direct writing may even make it possible to monitor a device's performance during the fabrication. In this article we present some schemes we recently developed for deposition of nanometer structures that may utilize the combined effects of the laser radiation, and the tunneling gap of a scanning tunneling microscope (see Figure 1).[9,10] The process is obviously highly nonlinear; it may involve several effects. These include excitation, ionization, heating, melting and evaporation by laser radiation or electrons, field ionization, chemical interactions, and material transfer between the tip, the sample, and the precursor gas. One or several of these effects may play a role depending on the type of application sought, however there is no complete control over the effects at this time.

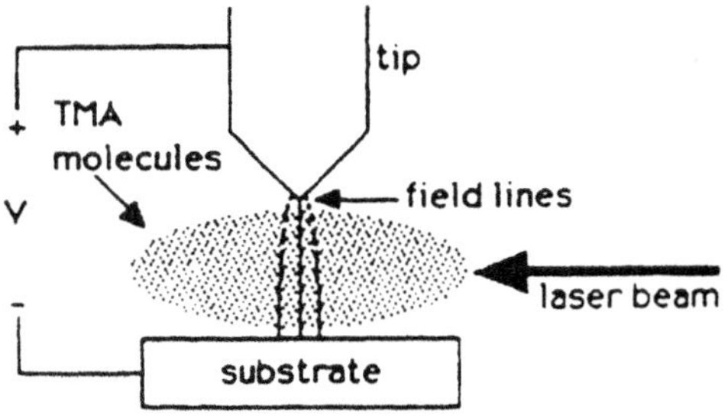

Figure 1. Schematic of the STM tunneling junction bathed in laser radiation.

FABRICATION OF RAISED STRUCTURES/MOUNDS

We have fabricated nanometer scale deposits of dimensions as small as 1 x 2.4 x 1 nm on highly ordered pyrolytic (HOPG) graphite surfaces using laser induced fragmentation and ionization of Trimethylaluminum [TMA, Al(CH3)3] in the surface-tip field of a scanning tunneling microscope.[9,10] The idea of deposition relies on a multiphoton process for fragmentation and ionization of TMA in the proximity of the gap of the STM. Radiation at 4300 Å (2.9 eV) from a tunable pulsed dye laser pumped by a 308 nm eximer laser bathes the tip-surface in a grazing angle configuration. It is known that a three-photon process using radiation of wavelength in the range 3,900-4,600 Å breaks the carbon-aluminum bonds, generating aluminum atoms and other radicals.[11] The wavelength of the radiation is short enough such that other subsequent multiphoton processes can occur such as ionization of the generated radicals, and Al atoms. Moreover, a competing process involves four photon ionizations of TMA, whose ionization potential is in the range 9-10 eV, yielding molecular ions. The ions are guided to the surface by the electric field between the STM tunneling tip and the surface. The nature and degree of the sticking depends on the materials of the deposits and the surface. Since the effective field of the tip is confined laterally to a few nanometers, the deposition is also controlled with such resolution. The structures deposited are subsequently imaged with the STM.

In addition to photoionization of TMA via a four photon process, the radiation of wavelength 4300Å (photon energy of 2.9 eV) can dissociate TMA and happens to be near (by design) the multiphoton resonances of the Al atom. Therefore, the laser can be tuned to certain resonances while still being able to photodissociate the TMA molecule. The laser with wavelength at 430 nm excites the freed Al atom from the ground state 3pllPJ by the absorption of two photons. The excited A1 atoms can further absorb a third photon, which ionizes them. It is seen that tunability can be used to enhance the deposition of Al atoms over other species, resulting in selective deposition. Moreover, TMA is chosen because at room temperature, the molecules are dimerised to form $A1_2(CH_3)_6$, which is a liquid with a vapor pressure of 8.4 torr. The high vapor pressure allows experiments to be performed at room temperature without major condensation problems. In addition, TMA has a relatively high pyrolytic decomposition temperature of 350° C, which eliminates the possibility of pyrolytic deposition when slightly elevated temperature study is needed, or when the laser beam grazes the surface. The sample is loaded and a vacuum of 10^{-9} torr is established. TMA is then introduced to a pressure of up to 10^{-4} torr via a gas handling system which allows for static fills or a flow mode.

We made a number of deposits using single laser pulses with a tunneling current of 1 nA and different tip biasing voltages, ranging from 0.8 to 3 V. Figure 2 shows a deposit made with one laser pulse and a tip bias of 1 V. The deposit appears near the middle of the image, which also shows the individual graphite atoms. The deposit appears to be composed of four or five aluminum atoms, since the laser wavelength was tuned to an aluminum resonance line, where the yield of aluminum ions is about six times more than the yield of other species. It should be mentioned that direct identification of the deposit is not yet possible. The line profile also shows some variation in the corrugation of the deposited atoms. Figure 3 shows another interesting deposit; it was made with one laser pulse and a tip bias of 1.1 V. Its area is approximately 14 Å x 13 Å and the individual atoms in the deposit are not resolved. The line profile through the deposit gives a corrugation height of 2.0 Å above the corrugation of the graphite atoms, compared to 0.6 Å for the previous case (Figure 2). Using 0.6 Å for the corrugation height of a single layer, this corresponds to about three atomic layers, or a total of approximately 20 atoms in the deposit.

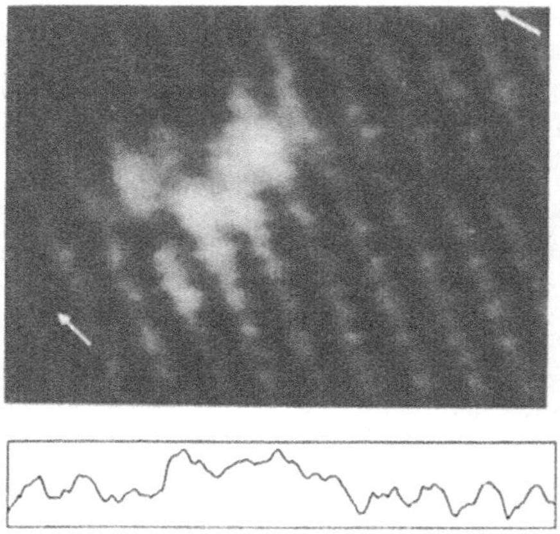

Figure 2. A deposit of TMA made with one laser pulse and a tip bias of 1V. The line profile through the deposit shows a corrugation height of approximately 0.6Å above the graphite atoms.

When the laser power is reduced, and the biasing voltage is increased, the field and current of the gap dominate and we tend not to break the molecules. Figure 4 shows one of our smallest deposits produced under these conditions namely a single molecule of TMA. The image (35 Å x 35 Å) was taken after the application of a 0.02s, 4V pulse to the tip, pointing to a doublet like shape indicative of a dimer that consists of two semilocalized three-center CA12 bonds. Other deposits we made also show the doublet structure of the molecule.

We now examine the possibility of erasing and its addressability. Figure 5(a) shows a graphite surface with several nanometer-sized structures. The tip was positioned directly above

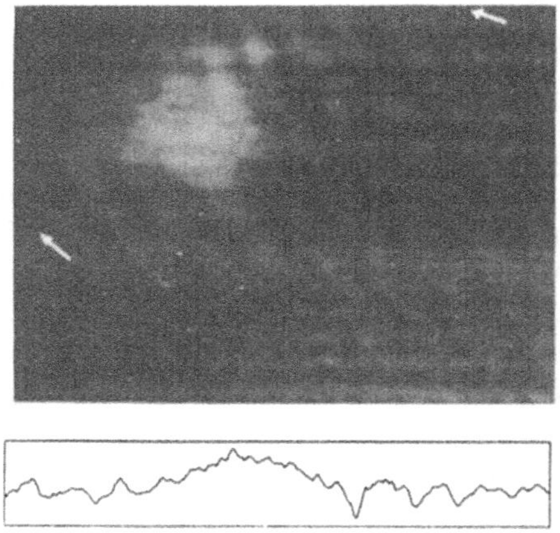

Figure 3. A deposit of TMA made with one laser pulse and a tip bias of 1.1V. The line profile through the deposit shows a corrugation height of approximately 2.0 Å above the graphite atoms.

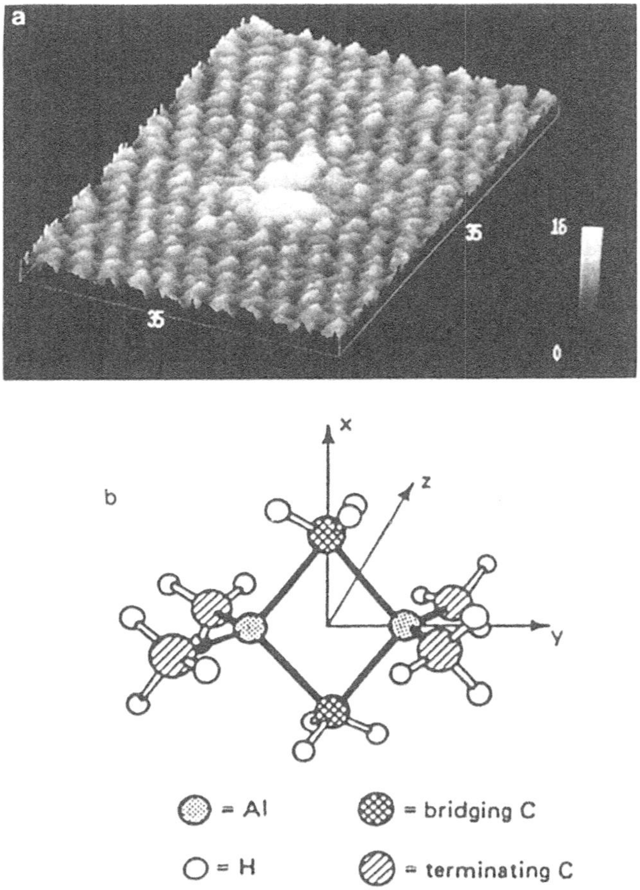

Figure 4. A deposit of a single TMA molecule on graphite (a), along with the schematic of the chemical structure of dimer (b).

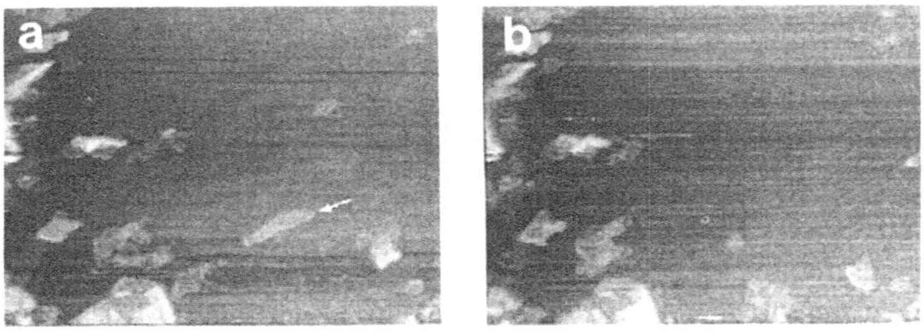

Figure 5. (a) STM image (255 nm x 255 nm) of a graphite surface, showing several deposits. The tip was placed above the deposit indicated by the arrow, and a laser pulse was activated. (b) STM image of the same area showing that the indicated structure has disappeared.

one of the structures (labeled by an arrow). The surface was then irradiated by a single pulse that filled the gap. Figure 5(b), taken right after, shows that this structure has disappeared while all of the other structures have not been affected. Since all of the structures got exposed to the laser radiation as the beam cross section was larger than the area of the sample, then it is clear that the laser in this technique can only affect the structure placed right under the tip. This spatial selectivity of the erasing scheme (addressability) constitutes the best resolution achieved so far and we believe it will have significant implications.

In another development, we have used our technique for depositing material in grooves and holes that have been prepared by other means.[10] We drill holes in a graphite surface by a high voltage pulse across the gap. The STM-laser technique is then used to fill the holes. To illustrate the power of our technique, we show in Figure 6 a series of images of the growth of the deposit that eventually fills the hole. Moreover, our measurements show that such fillings require higher laser intensities to erase than the deposits made on flat surfaces. This might add another dimension and feature to the versatility of the technique.

We also fabricated nanometer scale lithographic structures on silicon. Silicon is reactive and forms "contamination" layers of the precursor gas. In this case the process proceeds by electron polymerization and activation of the reaction of the gas as well as of the absorbed layer with the silicon surface. The substrates used were n-type silicon (111) samples with a conductivity of 0.1 ohm-cm.

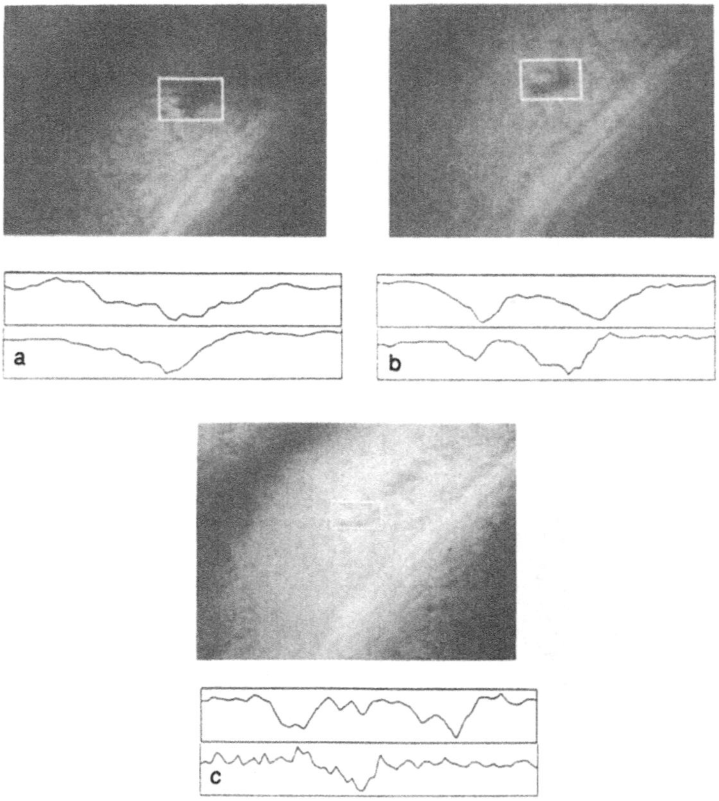

Figure 6. Using the laser technique to fill a previously drilled hole. (a) The hold with two line profiles across with a peak-to-valley height y of 84.2 Å. (b) and (c)The hole and growth of the deposit in the hole after 7 and 61 second laser irradiation, with peak -to-valley of 46 and 37 Å, respectively.

The samples were chemically prepared with the method of Ishizake and Shiraki[12] in which 49% HF solution is used in the finishing step to remove any oxide on the sample surface. An as-prepared sample was loaded into the STM, which was immediately loaded into a UHV chamber. The chamber was then pumped down to a pressure of 8 x 10^{-9} torr. The sample surface was first imaged with the STM operated in the constant current mode. Images of the surface show typical topography where residual surface adsorbates appear as clusters with an average size of 20 nm. The sample is then exposed to TMA at a pressure of 10^{-4} torr for a certain period of time to form an adlayer, and then the gas is pumped out or changed to a lower value. After positioning the tip over a certain point on the surface, and selecting the tip-surface spacing by establishing the appropriate tunneling current and biasing voltage, we apply a certain voltage pulse at the tip while holding the tip in place. The surface is then immediately scanned (no more than 20 s delay).

As an example we show in Figure 7 a three-dimensional view of a topographical mound that was made using a single pulse of 5 Volt amplitude, 0.2 second duration, and with 1 nA current. In the process, the tunneling electrons are used such as electron beam to expose the

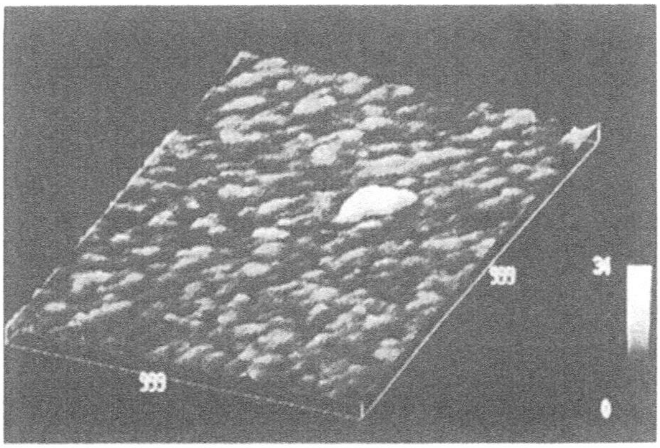

Figure 7. A three-dimensional STM image of a 12 x 14 x 16 nm mound on H-passivated silicon. Hydrocarbon species appear as clusters with an average size of 10 nm.

carbonaceous species adsorbed on the sample surface. The resultant raised features, therefore, are real topographic structures due to the build-up of the polymerized species.

It is also interesting to note that the process can be reversible. Our results indicate that complete or partial erasure of mounds could be made with additional processing. In the procedure we position the tip right over a structure that has been previously made, and apply a series of pulses. Figures 8 (a) and 8 (b) show that a single pulse erased the lower part of the structure with the upper part essentially preserved, thus effectively reducing the size of the structure. We find that the polymerization and adhesion to the substrate are strong enough so as not to be erased by single pulses. Usually multiple pulses can achieve complete erasure of

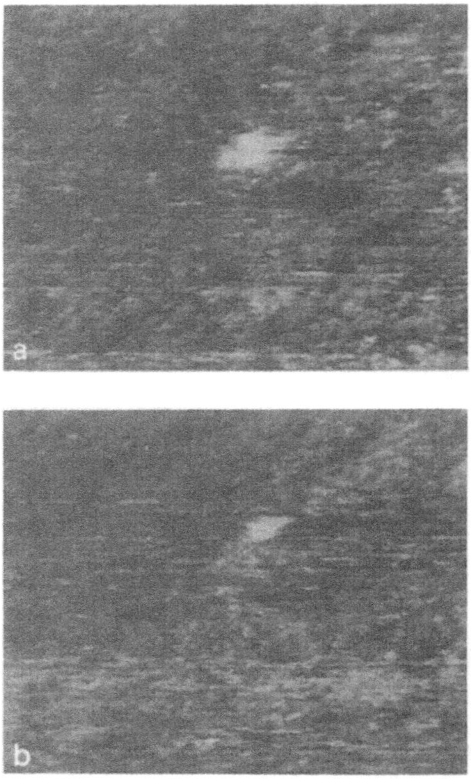

Figure 8. Demonstration of erasure with the STM. A single voltage pulse used on the deposit in (a) erased the lower part of the deposit, as shown in (b).

the structures. Additional fabrication processes at nearby sites, with distances longer than 10 nm, however, hardly affect the integrity of a given structure.

The next series of measurements involved continuous fabrication. For this purpose the tunneling current and bias were set to the appropriate values. The voltage across the gap was now increased to the required level for fabrication while simultaneously the tip was externally set in motion at a given speed. The motion and the high voltage were then simultaneously terminated at the end of the fabrication. Figure 9 shows a grid fabricated using this procedure. In Figure 10, we present a pattern made with this process. This pattern shows the ability of making arbitrary shapes.

Another procedure we employed for fabrication might be best described by the terms "shading". In this case a certain area of a surface was chosen and its coordinates (or boundary) fed into the computer. The process then proceeds by fabricating 256 close lines inside the area, effectively, shading the region. Figure 11 is an example of such fabrication on silicon where three boxes of 500 x 500 Å were shaded.

We refer the reader to Reference 10 for more details on the basic principles of the fabrication process on silicon. These include several issues such as efficiency, resolution, repeatability, reproducibility, dependence on the voltage, current, and duration of the pulses, tip-surface distance, conditions of the surface, adjacent processing, repeated processing and reversibility, etc.

Figure 9. A grid fabricated with the STM.

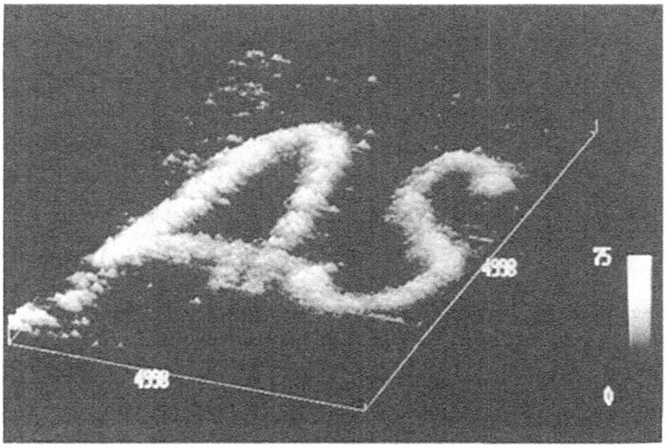

Figure 10. Ten patterns in the shape of alphabets are fabricated.

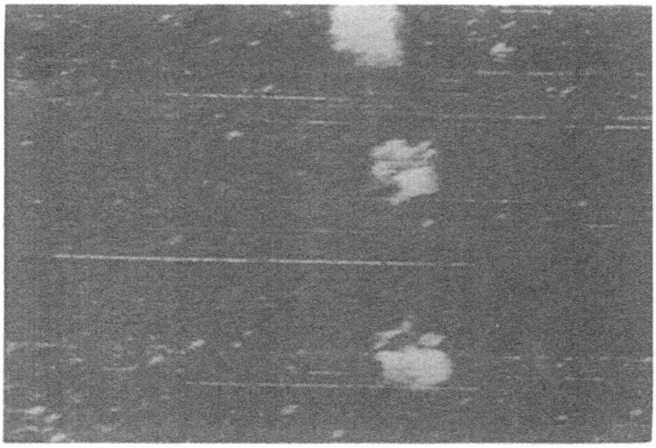

Figure 11. The three structures are fabricated by fabricating 256 close lines within the areas of each of the resulting structures.

STRONG FIELD EFFECTS ON THE FABRICATION

If the STM is to be used as an instrument for producing patterns with nanoscale line widths for semiconductor device applications, the fabricated lines should be uniform with well-defined edges. We investigated the uniformity of STM fabrications on the "conlamination resist" of wet chemically cleaned silicon. Our results showed that continuous fabrication produces isolated nanoscale mounds along the motion of the tip rather than uniform lines.[10] For example, Figure 12(a) shows a line that was fabricated with a single sweep of the tip at a bias voltage of 7 V and a tunneling current of 1 nA. The average tip speed was 100 Å/s. The line segment shown in the figure is ~ 100 nm long. It consists of 11 well-defined spots ranging in diameter from 4.2 to 10.8 nm and a standard deviation of 2.6 nm. The spot spacing is somewhat regular showing an average spacing of 8.4 nm. We emphasize that the dominant feature of all lines fabricated by a single sweep of the STM tip is their spotty nature. In an attempt to improve the quality of the fabricated lines, we experimented with producing wider patterns by sweeping the tip through many finely spaced lines. Figure 12(b) shows a pattern produced by sweeping the tip through 256-100 nm long, horizontal lines at an interval spacing of 0.4 nm, but again showing a spotty nature. These results are explained as a consequence of a self-limiting field effect in this fabrication regime. We have reason to believe that the structures are actually topographical ones made spotty by current tracking in which the electron beam current will remain directed towards the part of the sample which is closest to the tip. Therefore, when a raised structure is fabricated on the surface, the electron beam will remain directed toward this structure until the tip, has moved sufficiently far away so that the flat, unfabricated part of the surface is closer. The effect of current tracking is the creation of lines made of discreet mound, rather than continuous lines.

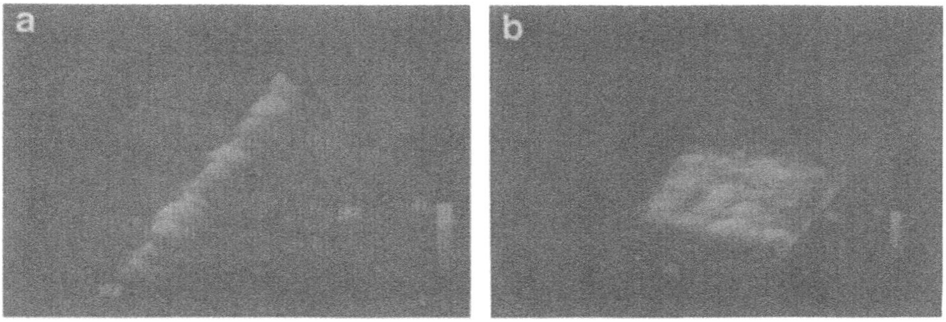

Figure 12. (a) A line fabricated with a single sweep showing the spotty nature of the fabrication. (b) Fabrication of 256 finely spaced lines showing spotty nature in two dimensions.

ETCHING/FABRICATION OF NANOGROOVES

More recently, the high degree of control on interdistances of STM has revealed another promising feature that is useful for surface modification, namely control of the chemical potential between the tip and the surface.[13] At small enough interseparations, the chemical forces acting on the surface become very strong, and even comparable to the binding forces involved. This chemical effect combined with the strong field effect were recently used to demonstrate the ability of the STM to remove atoms with atomic resolution, off a clean silicon surface, a surface that involves strong and covalently bonded atoms.[14] Specifically, by combining the strong field effect and the chemical force effect, silicon atoms and silicon clusters up to tens of atoms were reproducibly transferred from the surface to the tip using negative biasing at the tip. Moreover, the clusters removed were subsequently redeposited on other locations on the surface using opposite biasing.

Here we report on successful fabrication of nanoscale grooves on a silicon surface that has been prepared by chemical etching (Clery[1]). Micrometer-long continuous grooves that are as narrow as 50 Å full width at half maximum (FWHM) were produced with good uniformity on H-passivated silicon under ultrahigh vacuum conditions. Figure 13 shows a three-dimensional image of silicon surface showing a groove pattern etched using a tip bias of 400 mV and 1 nA tunneling current. The pattern consists of four straight segments meeting at right angles, each of 50 Å FWHM and 15Å depth. Figure 14 shows several line profiles across different parts of the groove to illustrate its quality. A closer look at the channels show some pile up of material at the sides. If one assumes near "local" redistribution of the material after it gets picked up by the tip, i.e. the material gets shaken off locally, then one concludes that the actual depth of the material that has been excavated is 9 Å. This points to the conclusion that there is no appreciable pile up of material on the tip, and as such there is no need for reversal of voltage biasing to shake it off as was required in the fabrication on clean silicon samples.[14]

Figure 13. A groove fabricated in silicon.

NANOFABRICATION ON PHOTORESIST-COATED SILICON

Recently techniques have been developed which use a scanning tunneling microscope (STM) as a tool to fabricate nanometer-scale structures on electron beam (e-beam) resist. [15-18] The main motive for using e-beam resist in the nanometer-scale fabrication of structures is the potential for pattern transfer, since resists exposed to electrons can be developed and used as masks for further processing. Although other techniques already exist which are also being used for the fabrication of structures in the nanometer range, the most successful of which is electron beam lithography, it is hoped that STM-based techniques will be able to complete and even surpass the capabilities of these more conventional techniques. In fact, Dobisz and Marrian[16] have already shown that nanofabrication with an STM on resist is capable of outperforming

Figure 14. Line profiles across different parts of the groove of Figure 13.

standard electron beam lithography in both the achievable resolution and the dimensions of the fabricated structures. Although in these studies the STM images of the areas that had been fabricated on did not show changes in the surface due to the fabrication process, the patterns were observable by scanning electron microscopy (SEM) after the sample was chemically developed. Line widths as small as 27 nm and line spacings as small as 55 nm as measured using an SEM were reported in contrast to minimum line widths of 95 nm fabricated using a 17 nm (1/e diameter) 50 kV electron beam.[16] Although these studies have shown tremendous potential, it was noted that this achievement may not have been the ultimate capability of this technique.

Recently we reported on STM studies that have been initiated in our laboratory to improve the resolution and to study the systematics of the lithographic process, such as dependence of line or groove fabrication on the biasing voltage, the tunneling current, and the tip speed (and hence the tip sample separation and the line dose).[19] In this study we only focussed on the STM exposure process and not on pattern transfer. Extensive studies on the development of STM exposed resist using the same resist as in this work have been described.[16] Our results show that the process is nonlinear, switching over from producing structures that appear as mounds to structures that appear as grooves as the fabrication voltage drops below approximately 6 V. Structures like mounds of width as small as 18 nm and with a corrugation as small as 2 nm and structures like grooves of widths as small as 15 nm and depths as small as 0.5 nm were fabricated.

The sample used in this experiment was (111)-oriented p-type silicon with a resistivity between 0.01 and 0.02 ohm-cm. It underwent a routine initial cleaning with acetone, isopropyl alcohol (IPA), and deionized water. The silicon sample was then dipped in buffered hydrofluoric acid to remove any oxide present on the surface. A second routine cleaning with acetone and IPA was performed shortly before the resist was applied. The resist used was SAL 601 manufactured by Shipley. It was spun onto the sample at a rate of 8000 rpm giving a thickness of 37.5 - 40.0 nm as measured by profilometry. The resist-coated sample was then baked at 70°C for 30 minutes.

The biasing configuration of the microscope is such that the tip is always grounded while the sample is positively biased; therefore, electrons flow from the tip to the surface. This particular biasing configuration is specific to our instrument and cannot be changed. The surface was scanned before and after the fabrication process using the same biasing configuration (i.e., grounded tip, positively biased sample). Biasing and current conditions giving stable images of both the pre- and postexposed resist occurred with voltages between 7 and 11 V and with currents between 0.10 and 0.12 nA.

For fabrication where the tip is initially idle and using constant biasing voltages larger than 20 V, the process failed to produce localized patterns, but rather altered the entire scan area, producing a much more corrugated surface. Presumably, this was due to the large exposure the surface received while the tip was idle at high bias before the line scan was initiated. However, as the biasing voltage was reduced to 15 V, the surface modifications become more controlled and localized showing faint patterns that look like raised structures along the line which the tip moved. As the biasing voltage is dropped further, e.g., to 10 V, the process becomes more consistent for a range of tunneling currents (0.1-0.5 nA). Typically we get structures whose widths are in the range of 20 - 30 nm, and whose corrugation are in the range of 1.0 - 1.5 nm. Figure 15(a) shows an example of these structures, and Figure 15(b) gives some line profiles displaying the width and the depth of these structures, showing a height plot averaged over the length of the line. Generally speaking, the height was of the order of 0.5 to 1.0 nm and the FWHM of the order of 20 nm for the majority of the fabricated structures in this image. From these studies on the fabrication of mounds, we conclude that the size of the structures decreases as the voltage decreases up to a certain limiting value which is possibly related to the activation or minimum exposure energy of the photoresist. Further investigation is necessary to determine the activation voltage. We have reason to believe that the structures which appear as mounds might actually be topographical since they exhibit the discontinuity which is typical of current tracking, a phenomenon that we mentioned earlier. For voltages below 6 V, the process becomes ineffective. However, if the tunneling current is increased to a level of 1 nA at these lower voltages, thus bringing the tip closer to the surface the tip starts to remove material or to plow into the surface, producing trenches or grooves. Tip-sample contact similar to this has been reported by other researchers.[1,4] Typical results of groove

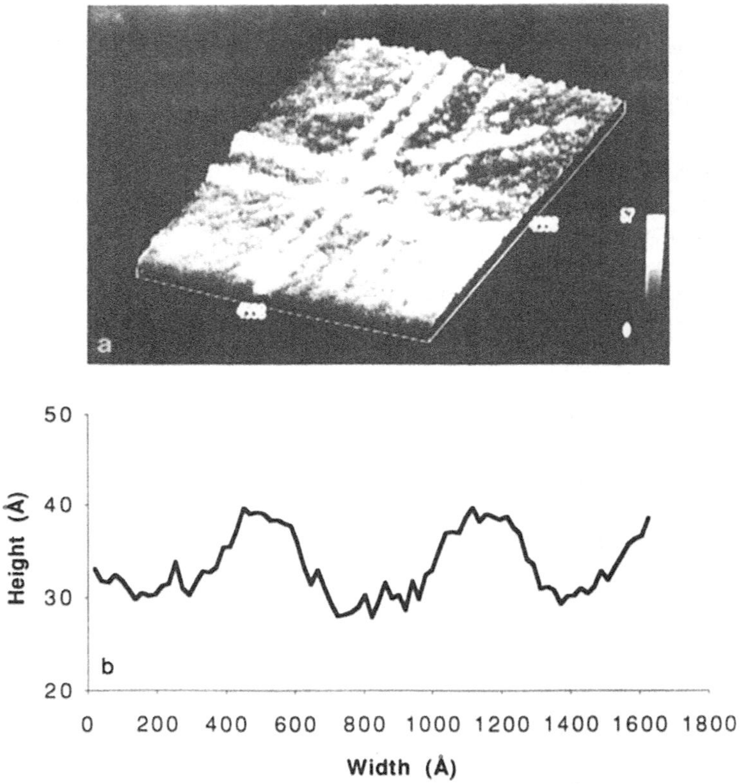

Figure 15. (a) Mound lines fabricated on photoresist-coated silicon. (b) Line profiles across the lines.

fabrication are shown in Figures 16(a) and 16(b), and the corresponding line profiles shown in Figures 16(c) and 16(d) were calculated in the same way as the line profile for Figure 15(b).

Although the depth of the grooves is comparable to the height of the mounds (approximately 1.5 nm), the FWEIM is measurably larger, averaging between 30.0 nm and 40.0 nm. The process gives the narrowest grooves of 15.0 nm width at a biasing voltage of 5 V; but these grooves are quite shallow with a depth of about 0.5 nm. At lower voltages but the same current, namely 3, 2, and 1 V, the process produces wider grooves, (about 30 nm), with larger depths, (about 1.5 nm). For even lower voltages, the tip starts to make appreciable mechanical contact with the resist, essentially plowing material to produce grooves that are very wide.Fabrication at higher voltages in the range of 35 - 50 V was observable only when the voltage was raised to these levels after the tip was set in motion to minimize the line dose while the tip was idle at the initial starting point. In this case raised structures 70.0 - 80.0 nm in width (at its maximum width) and 1.5 - 2.0 nm in height were made when the biasing voltage was ramped from 7 to 50 V and back to 7 V over the span of a single line scan. The raised structures do not appear entirely over the line scan but rather at its central section. Other trials at a lower maximum biasing voltage of 35 V yielded lines with a maximum width of 47.0 nm and a height of 1.0 nm.

We examined the relationship between the dimensions of the fabricated structures and the biasing voltage, tunneling current, and speed of scan (hence tip-sample separation and line dose). An analysis was done on the dimensions of the fabricated structures as a function of the line dose, which is the ratio of the tunneling current to the tip speed. The results indicate that

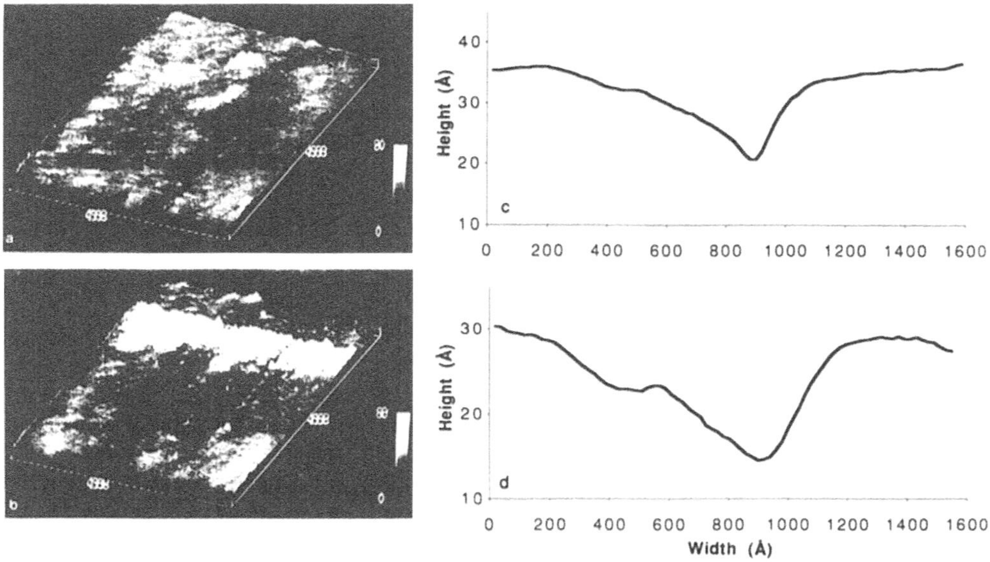

Figure 16. (a) and (b) groove lines fabricated on photoresist-coated silicon with line profiles across them given in (c) and (d), respectively.

in this range (80 - 2000 mC/cm), the dimensions do not change appreciably with the line dose. The fabrication process, on the other hand, depends strongly on the biasing voltage. Our results show that in the regime in which the line dose does not have a strong effect on the dimensions of the fabricated structures (80 - 2000 mC/cm), the strongest influence on the size of these structures is the tip-sample voltage, rather than the tip-sample current. A change by a factor of 5 in the current (from 0. 1 to 0.5 nA) while the voltage was kept constant resulted in an increase in the FWHM of the fabricated structures by approximately a factor of 1.2, from 20 to 24 nm; whereas a change by a factor of 5 in the voltage (from 10 to 50 V) while the current was kept constant, increased the FWHM of the structures approximately by a factor of 4, from 20 to 83 nm.

PARALLEL FABRICATION

STM-based fabrication provides a tremendous refinement in writing resolution, however it is inherently slow as it requires the movement of a mass, namely the tip. Since the objective of commercial application is the development of a high speed of access and/or writing while maintaining high precision, and mass production of a few thousand devices at the same time, a multitip linear array arrangement or "comb" may alleviate these obstacles and hence turn the concept of writing with atoms and molecules into a device of suitable commercial application. The linear array writer is a closely packed, equally spaced set of needles with spacing as small as a few micrometers.

We tested this concept by preparing an array of two and three tips. These multipeak tips were produced *in situ* by zapping a single tip with a large current pulse. The resulting peak configuration depends on the magnitude and duration of the current.[10] We used such multi-peak tips to fabricate more than one mound by a single fabrication process on silicon. Figure 17 shows several deposits of this kind each showing a substructure of three mounds of the same interspacing. It appears that these were made by three sharp intrusions on the tip, separated from each other by approximately 100 Å.

Figure 18 shows two parallel lines made on the surface by moving the tip diagonally. Figure 19 shows an example of a pattern fabricated with a two-peak tip for further illustration. In the vertical direction we see double lines, while in the horizontal direction, we see a single line. One can also see that the double lines merge into one when the tip motion changes from the vertical to the horizontal direction. This exercise also shows that the structure made by the front peak does not get compromised or erased by the back peak as it cuts through it.

STABILITY AT HIGHER TEMPERATURE

We are in the process of integrating the fabrication capability with novel molecular beam epitaxy (MBE) methods to fabricate and analyze two and three-dimensional nanometer scale structures such as quantum wires and dots, quantum gratings, arrays of quantum dots etc. We are presently using these techniques to construct and test the quantum interference transistor, a micrometer size MOSFET with a nanometer scale grating or grid embedded in its gate area. These advances have important implications to mass storage of information which may lead to great reductions in the sizes of electronic circuits and devices.

We tested the suitability of the structures in this regards by examining their stability at higher temperatures via heating the substrate and the nano structure itself by a continuous wave laser beam. The laser beam was directed at the structure obliquely with an incident angle of 70 degrees. It appears that for up to 25 minutes of heating, which we believe might have raised the temperature to 200° C - 250° C, the structure remained stable except for loss of

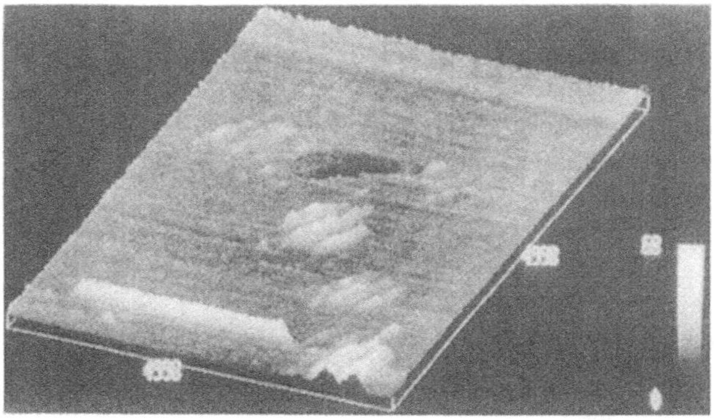

Figure 17. An image of a silicon surface showing triplets of structures made in single trials.

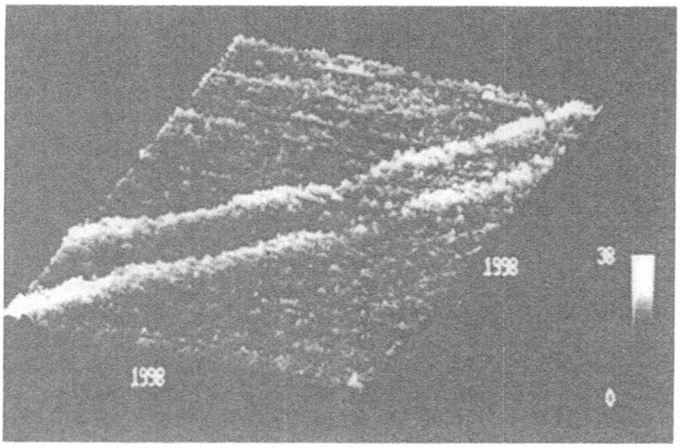

Figure 18. Two parallel lines made by a two-peak tip.

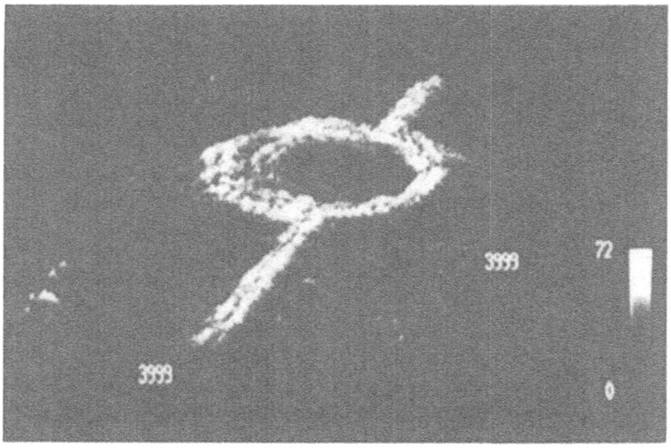

Figure 19. A pattern fabricated by a two-peak tip.

some periphery. On the other hand, the continuous heating started to affect the contamination layer in the unprocessed region of the silicon sample long before it did anything to the structure itself. Studies at higher temperatures of several hundreds degrees are underway.

ACKNOWLEGMENTS

This work was supported by the Naval Research Laboratory under contract NOS. N00014 - 87 - K - 0354, N00014 - 90 - J -1004 and N00014 - 92 - J - 1519.

REFERENCES

1. See, for example, the news report by Daniel Clery, Nanotechnology rules, OK!, *New Scientist* 133:1811, 42-46 (1992).
2. See various articles in *IBM J. Res. Develop*. 30 (1985).

3. E. E. Ehrichs, S. Yoon, and A. L. de Lozanne, Direct writing of 10 nm features with the scanning tunneling microscope, *Appl. Phys. Lett.*, 53, 2287-2289 (1988).
4. J. A. Dagata, J. Schneir, H. H. Harary, C. J. Evans, M. T. Postek, and J. Bennett, Modification of hydrogen-passivated silicon by a scanning tunneling microscope operating in air, *Appl. Phys. Lett.*, 56, 2001-2003 (1990).
5. S. T. Yau and M. H. Nayfeh, Nanolithography of chemically prepared si with a scanning tunneling microscope, *Appl. Phys. Lett.* 59, 2457 (1991); M.H. Nayreh, Fabriction of nonometer scale structures, SPIE Institutes, IS 10, 200-217 (1993)
6. D. M. Eigler and E. K. Schweizer, Positioning single atoms with a scanning tunneling microscope tip, *Nature* 344, 524-526 (1990).
7. H. J. Mamin, R. J. Hamers, and D. Rugar, Atomic emission from gold scanning-tunneling-microscope tip, *Phys. Rev. Lett.* 65, 2418-2421 (1990); I.-W. Lyo and P. Avouris, Field induced nanometer-to-atomic-scale manipulation of silicon surfaces with the STM, *Science* 253, 173-176 (1991).
8. See, for example, R. M. Osgood and T. F. Deutsch, Laser-induced chemistry for microeclectronics, *Science* 227, 709-714 (1985); D. J. Ehrlich and J. Y. Tsao, Nonreciprocal laser-micromechanical processing: spatial resolution limits and demonstration of 0.2 micrometer linewidths, *Appl. Phys. Lett.* 44, 267-269 (1984).
9. S. T. Yau, D. Saltz, and M. H. Nayfeh, Laser-assisted deposition of nanometer structures using a scanning tunneling microscope, *Appl. Phys. Lett.*, 57, 2913-2915 (1990); S. T. Yau, D. Saltz, A. Wriekat, and M. H. Nayfeh, Nanofabriction with a scanning tunneling microscope, *J. Appl. Phys.* 69, 2970-2974 (1991); S. T. Yau, D. Saltz, and M. H. Nayfeh, Scanning tunneling microscope-laser fabrication of nanostructures, *J. Vac. Sci. Technol.* B9, 1371-1375 (1991).
10. X. Zheng, S. T. Yau, and M. H. Nayfeh, *Parallel fabrication on chemically etched silicon using scanning tunneling microscopy, Ultramicroscopy* 42-44, 1303 (1992); J Hetrick, X. Zheng and M.H. Nayfeh, Strong field effect in nanofabrication on chemically prepared silicon, *J. Appl. Phys.* 73, 47221-4723 (1993).
11. S. A. Mitchell and P. A. Hackett, Pulsed visible laser photolysis of B(C2H5)3, A12(CH3)6, Ga(CH3)3, and In(CH3)3: Multiphoton ionization spectra of A1, Ga, and In atoms, *J. Chem. Phys.*79, 4815-4822 (1983).
12. A. Ishizaka and Y. Shiraki, Low temperatuer surface cleaning of silicon and its application to silicon MBE, *J. Electroshem. Soc.* 133, 666-671 (1986).
13. N. D. Lang, Apparent barrier heights in scanning tunneling microscopy, *Phys. Rev.* B 37, 1 10395-10398 (1988); S. Ciraci and E. Tekman, Theory from transition from the tunneling regime to point contract in scanning tunneling microscopy, Phys. Rev. B 40, 11969-11972 (1989); J. K. Gimzewski and R. Miller, Transition from the tunneling regime to point contact studied using scanning tunneling microscopy, *Phys. Rev.* B 36, 1284-1287 (1987).
14. I. W. Lyo and P. Avouris, Field-induced nonometer-to-atomic scale manipulation of silicon surfaces with the STM, *Science* 253, 173-176 (1991).

A MICROSCOPY FOR OUR TIME

Elinor Solit

The Cambrex Group
Microscope Technology & News and The Microscope Book
Boston, Massachusetts 02109

Abstract: One of the newest and most revealing microscopies is designed around a lensless system and a probe tip. With patents less than 15 years old, and a vigorous commercial marketplace, applications are springing up swiftly. This paper describes the colorful events behind the development of probe microscopy and weighs the science and the circumstances which are likely to affect it.

Microscopy is a business full of enterprise, luck and coincidence. We rarely have the opportunity to watch it unfold. The true origins of light microscopy for instance, are lost in the mysteries of the 17th century. And if we *had been* able to watch it, it would have taken centuries to measure growth.

Probe microscopy may also have its beginnings in the work of some forgotten physicist, but the action part is occurring right before our eyes. Nanotechnology is the microscopy of our time. So I hope it will be of some interest to trace its progress and the role that the business side plays in bringing science into the lab.

The first modern patent describing a scanning probe microscope was filed by Binnig and Rohrer, two IBM employees, less than 15 years ago. That was 1980. At that time every aspect of microscope technology used lenses of some kind, whether they were light microscopes with fine glass lenses or electron microscopes utilizing magnetic fields to form images. We all knew that resolution was limited by the laws of physics. Binnig and Rohrer, liberated from the constraints of lenses, described a lensless instrument that could visualize materials at the atomic level.

In 1986, Binnig and Rohrer shared one half of the Nobel Prize for Physics. The other half of the award went to Ernst Ruska for the first electron microscope. It is interesting that the Nobel judges rated the STM at the same level of significance as the invention of electron microscopy.

The commercial side of microscopy, the supplier base, is what creates the market and fosters its applications. The scientist may "father" the market by seeding it with the inspiration for the technology, but the supplier is definitely the "mother" that incubates the product into existence. If commercial companies cannot foresee a profitable response to a new instrument, it will not be developed. Thus the chain of events that produces new instrumentation, followed

Atomic Force Microscopy/Scanning Tunneling Microscopy 2
Edited by S.H. Cohen and M.L. Lightbody, Plenum Press, New York, 1997

by even newer science will be broken. We must have suppliers who are imaginative enough to visualize a new product or technology, who are willing and able to take reasonable risks, and close enough to the market to evaluate applications and support their growth.

If we look for the commercial "driver" in the field of probe microscopy, Digital Instruments comes to mind. Based in Santa Barbara, California, they can claim the largest segment of the installed base as it exists today. The company was founded and is still run by Virgil Elings. An instructor on the staff of UCal, Santa Barbara, Elings always operated a small company as a side occupation. In 1987, he had just sold one of them and the sale money was burning a $50,000 hole in his pocket.

Sabbatical time came around. Eling thought about his year off and decided it would be "fun to see atoms". He called Gus Gurley, a former student that he had not seen for 15 years, and offered him a job for a year. Gurley agreed. One year, $100,000, and a lot of learning later, Digital had their first instrument, an analog device. They decided to spend $1000 and place an ad in "Physics Today". One month after the ad appeared, Digital Instruments had sold enough product to recoup their entire investment.

Elings is the classical entrepreneur; his philosophy is that "money is not the product, it is the by-product". That is one of the principles that built an annual $20,000,000 business for Digital Instruments and until recently, they had the market to themselves.

But we must note a small company called Quan-Scan in Pasadena, California, and which in July of 1990, was granted a patent for a Scanning Micromechanical Probe Control System. The patent described a probe that at 40 nanometers could scan at a speed of 100,000 nanometers per second. Quan-Scan calculated that this was comparable to flying a 40 foot aircraft at a speed of 68,000 miles per hour while operating at an altitude of 10 feet and avoiding collisions with obstacles 1000 feet high.

One month later, the venture capital backers of Quan-Scan had withdrawn their support. However, some members of the Board had a vision of where the technology could go and were not ready to give up. Among them were David Hoyt, David Nelson and Jack Finnegan of the Finnegan Corporation a leading manufacturer of mass spec systems. Within six months, their finances had been reorganized and the company we know as Topometrix was born. Their approach has included a high-energy, high-profile marketing position, user-friendly software, and the formation of partnerships both technical and marketing. Their product line now includes the first commercial near-field microscope.

Companies with funding from venture capitalistic sources at the Topometrix level, that is several million dollars, usually try to go public within five to seven years so that may be what we will see from Topometrix.

There are other recognizable names in the field. Leica, for instance, is one of the largest microscope companies in the world. While Leica was consolidating itself out of several microscope companies, discussions were going on between Leica and Wyko, a Tucson, Arizona company specializing in surface metrology. Wyko, by the way, was founded by James Wyatt and two or three other University of Arizona scientists. The name Wyko was born when the founders, acting as consultants to IBM, decided to bill the company for work they had done. When, to their gratification, a check arrived from IBM, they needed a name in order to cash it, so they invented the word Wyko out of their last names. Last year, a marketing agreement was signed between Leica and Wyko, and now Wyko's two probe microscopes are in the Leica line.

And just last week, Carl Zeiss introduced a scanning tunneling microscope called The Beetle. The first probe microscope to carry the Zeiss name, its technology is licensed from Besoki Delta GMBH, a German company. The instrument is a stand-alone, and the price tag, says Zeiss, is $80,000. We will see it first at MAS in New Orleans in August.

Smaller companies in the field include Park Scientific Instruments - the creation of Sung Park and Sang Il Park, who were students of Calvin Quate at Stanford University. Inspired by work on ATM, in 1988 they decided to start a probe company. Perhaps because of the research orientation of the founders, Park's contribution to the technology emphasizes innovation.

Other participants in the probe market include Burleigh Instruments. Burleigh's original product line was laser interferometers. The resources required to make the products include mechanical skills and electro-optical expertise. David Farrell, president of Burleigh, sensed that these talents could be used to manufacture probes. Their low-end instructional and educational probe microscopes should broaden the marketplace, just as the ISI desk-top SEMs did with scanning electron microscopy in the 1970's.

Markets are brought into being by innovators, broadened by worthy competitors assuming they have some substance to offer, and extended by low-end producers. The six companies I've described, Digital, Topometrix, Leica, Zeiss, Burleigh and Park Instruments, all contribute some force of their own to the marketplace. These companies are certainly not the entire supplier base and I apologize to many that I've left out because of time constraints.

Who will dominate the probe business? The answer will depend on several factors. One is the speed with which the market develops. Another is the development of patents for new technologies. Then there is the financial and management strength of individual companies to consider. The direction in which applications move will be critical as well. And last but not least is the company itself. Entrepreneurial companies are different from venture-backed companies. Large companies are different from small ones. Old companies are different from new ones. So, much will be determined by industrial factors themselves. We need only look at the recent history of light microscopy to see how remarkable the commercial side of microscopy can become.

Now to the marketplace itself. Today's probe microscope base extends worldwide and is made up of about 2000 instruments. New sales account for about 350-400 instruments annually. We estimate the annual value in 1994$ to be between $50-60,000,000.

The product phase of the market consists of two segments - the primary market and the after-market. The primary market consists of hardware and software, some of which you will see here today. After-market items are tips, anti-vibration tables, video and hard copy systems. The approaches to these two market phases are quite different. Selling into the primary market is a lot of hard work - much plowing has to be done before the seed comes up, but selling into the after-market is like harvesting. You only need to be there when the fruit drops.

Where is probe going? Originally conceived as a tool for material scientists, it does not require magic powers to realize that the design of new materials will have a major impact on the lives of all of us. As cheap labor markets get used up, there is a race for materials which can lower product costs. The surface structure and behavior of new materials are major factors in their development and selection. Probe microscopy can be expected to contribute significant value to the materials market and to experience healthy growth in it.

In the life sciences, new ATM techniques have made probe useful for biologists and experimental scientists. Technology in this field is progressing at a pace we have never known before. The manipulation of live tissue is fact not fancy and new possibilities present themselves with mind-boggling speed. Virgil Elings told us that in two years, he expects 33% of his business to be generated by the life sciences. Growth in that sector, however, may well be affected in the short term by federal budgets. Government officials talk the growth, but walk the budget.

Even at this early stage, the commercial profile offers something for everyone. Today there are low-end microscopes, high-end microscopes, integrated microscopes, and dedicated microscopes. A joint venture between Topometrix and Hitachi in 1993 was significant. In a short period of time, it has produced an AFM device which can be used with Hitachi SEMs. This puts probe into the category of the after-market, with an installed base potential of thousands of existing instruments. Digital Instruments now markets an instrument which uses light microscopy as a locater for the probe operation. But should the configuration go the other way, and probe become the addition to the light microscope, the potential market jumps to the hundreds of thousands, even possibly into the millions.

So today, an individual wishing to look at samples at the atomic level has many choices. I would like to summarize with two comments. One is that as buyer and seller, remain flexible. The future of probe appears to be one with the sky as the limit, so leave yourself in a position to benefit from what might come. The other is that the source of the information presented here comes from the industry newsletter, *Microscope Technology & News*. Anyone wishing to know more about the subject is directed to call us at our Boston office.

SCANNING TUNNELING MICROSCOPY OF CHEMICAL VAPOR DEPOSITION DIAMOND FILM GROWTH ON HIGHLY ORIENTED PYROLYTIC GRAPHITE AND SILICON

A.F. Aviles,[1] R.E. Stallcup,[1] W. Rivera,[1,2] L.M. Villarreal,[1] and J.M. Perez[1]

[1]Department of Physics
University of North Texas
Denton, Texas 76203
[2]University del Cauca
Popayan, Colombia

Abstract: We report scanning tunneling microscopy (STM) studies of chemical vapor deposition (CVD) diamond film growth on highly oriented pyrolytic graphite (HOPG) and Si. The films were grown using hot-tungsten filament CVD. Using conditions typical for CVD diamond growth, we find that HOPG is etched by atomic hydrogen such that oriented hexagonal pits 50-5,000 Å in diameter are produced on the surface. Diamond crystallites are observed to nucleate on the walls of these pits and not on the smooth sp^2 bonded parts of the surface. At lower sample temperatures, HOPG is etched such that large circular pits approximately 10,000 Å in diameter and 7 Å deep are produced. Nanoscale linear structures, which we conjecture are hydrocarbon chains, are observed in these pits. These structures orient themselves when a voltage of 10 V is applied to the tip. The initial stages of diamond film growth on Si were studied. Polycrystalline films on Si approximately 2 μ thick were imaged in air from a micron to atomic resolution scale. The micrometer scale images show that these films consist of diamond crystallites with (100) or (111) oriented faces. Atomic resolution images of the (100) surface in air showed a 2 x 1 dimer reconstruction with a distance between dimer rows of approximately 5.1 Å.

INTRODUCTION

Chemical vapor deposition (CVD) growth of diamond films has recently attracted considerable interest due to the large number of potential applications of this material.[1,2] Polycrystalline film growth on substrates such as graphite, Si, Ni, and tungsten and epitaxial growth on diamond (100), (110), and (111) substrates have been achieved.[3,4] One of the goals of current research is growth of epitaxial or near-epitaxial films on nondiamond substrates. These attempts are hampered, however, by lack of knowledge about diamond nucleation sites and growth mechanisms. This dearth of information is due in part to the fact that diamond

Atomic Force Microscopy/Scanning Tunneling Microscopy 2
Edited by S.H. Cohen and M.L. Lightbody, Plenum Press, New York, 1997

growth involves unknown chemical reactions between hydrogen and sp^2 and sp^3 bonded carbon in the gas above the surface and on the surface. In this paper, we report scanning tunneling microscopy (STM) studies of hot-tungsten filament CVD diamond growth on HOPG and Si. HOPG has a simple atomic structure consisting of layers of sp^2 bonded carbon atoms with hexagonal symmetry. Using conditions typical for diamond growth, we find that HOPG is etched by atomic hydrogen such that oriented hexagonal pits 50-5000 Å in diameter are produced on the surface. Diamond crystallites are observed to nucleate on the walls of these etch pits and not on the smooth sp^2 bonded parts of the surface. At lower sample temperatures, HOPG is etched such that large circular pits approximately 10,000 Å in diameter and 7 Å deep are produced. Nanoscale linear structures observed in these etch pits orient themselves when a voltage of 10 V is applied to the tip. We conjecture that these structures are hydrocarbon chains formed as a result of a chemical reaction between atomic hydrogen and the HOPG substrate.

Polycrystalline diamond films on Si approximately 2 μ thick were imaged from a micron to atomic resolution scale. The micrometer scale images show that the films consist of diamond crystallites with (100) and (111) oriented faces. Atomic resolution imaging of the (100) faces was possible in air showing a 2 x 1 dimer reconstruction of the surface from which a distance between dimer rows of 5.1 Å was measured.

Our diamond growth system consists of a water-cooled hot-tungsten filament growth chamber attached to a UHV STM system. The diamond growth system allows control of gas flow rates, sample temperature, filament temperature, and growth chamber pressure. Our scanning tunneling microscope was manufactured by Burleigh Instruments, Inc.[5] Typical tip-sample voltages and tunneling currents used for imaging HOPG were 75 mV and 1.0 nA, respectively. We also have a Raman spectroscopy system set up adjacent to the UHV STM system for determining the diamond and graphite content of diamond films while they are in the growth or UHV STM chambers. The Raman spectroscopy system consists of an argon ion laser, double monochromator, and GaAs cooled photomultiplier tube. Raman spectra were typically taken using the 4180 or 5145 Å laser lines.

Figure 1(a) shows an STM image of an HOPG surface after approximately 5 minutes of exposure to atomic hydrogen using a filament temperature of 2100° C, sample temperature of 800° C, hydrogen gas flow rate of 200 sccm, and chamber pressure of 35 torr. These conditions are similar to those used for diamond growth except diamond growth uses in addition 0.5% of methane to hydrogen gas. Figure 1(a) shows oriented hexagonal etch pits with diameters from 50-5000 Å and depths of up to 1000 Å. Figure 1(b) shows an STM image of an HOPG surface after approximately 30 minutes of CVD diamond growth using the same conditions as in Figure 1 (a) with the addition of 0.5% methane. Crystallites are observed to have nucleated on the walls of the two hexagonal etch pits at the right-hand-side of the figure. Figure 2(a) shows a higher-resolution STM image of the crystallite on the wall of the top-most pit in Fig. 1(b). The crystallite, which has a diameter of approximately 100 Å, has nucleated between two atomic layers of the HOPG, as observed in Figure 2(a). We identify this crystallite as diamond based on the angles between the crystallite faces, which are inconsistent with HOPG, and Raman spectroscopy of the sample which showed a sharp diamond peak at some locations. Scanning electron microscopy of the sample also showed crystallites with shapes consistent with diamond. Figure 2(b) shows a larger area STM image of the same sample showing many hexagonal pits with crystallites on the walls. All of the pits are oriented and crystallites are observed only on the walls of the pits and not on the smooth sp^2 bonded parts of the surface. If the crystallites are oriented with respect to the walls of the pits, then since the pits are oriented with respect to one another, a continuous diamond film grown under these conditions may be near-epitaxial.

To investigate the reason for the hexagonal symmetry of the etch pits in Figure 1(a), HOPG was exposed to atomic hydrogen using the same experimental conditions as in

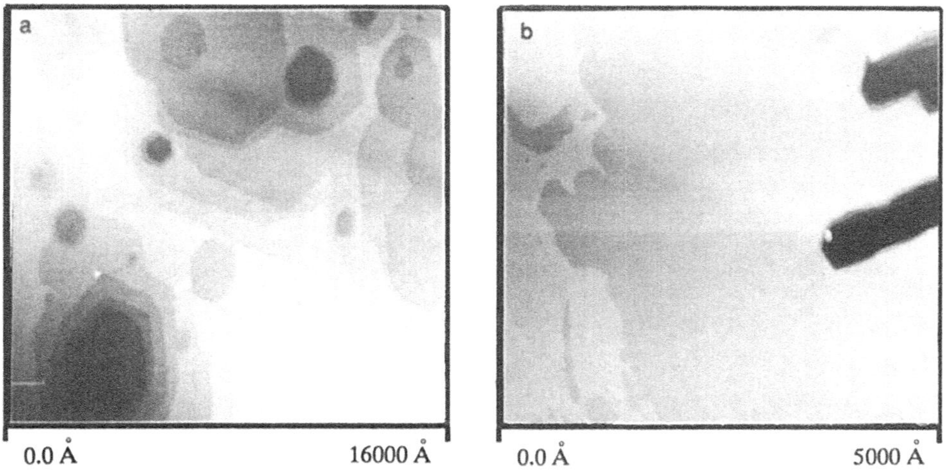

Figure 1. (a) HOPG graphite after 5 minutes of exposure to atomic hydrogen using a filament temperature of 2100° C, sample temperature of 800° C, hydrogen flow rate of 200 sccm, and chamber pressure of 35 torr. Oriented hexagonal etch pits are observed. (b) HOPG after 30 minutes of growth using typical CVD growth conditions as in (a) with the addition of 0.5% methane. Diamond crystallites are observed to have nucleated on the walls of the two etch pits on the right-hand-side of the figure.

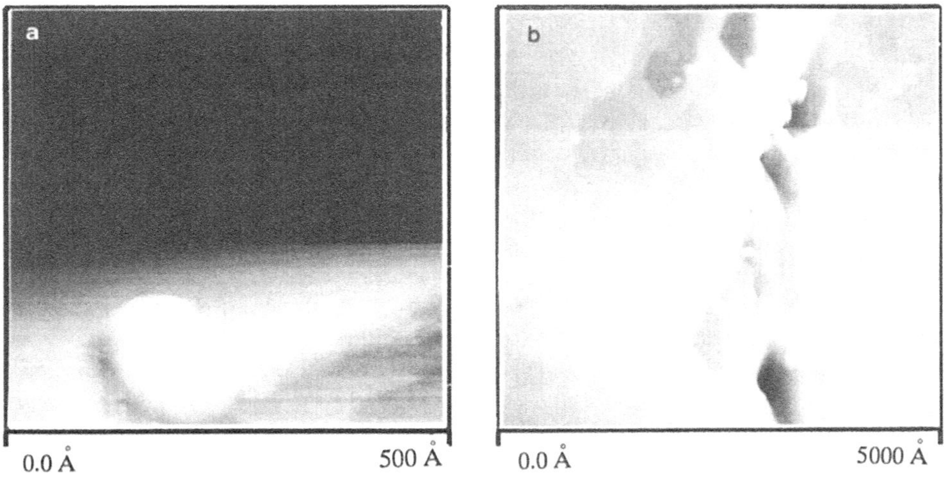

Figure 2. (a) High-resolution STM image of the crystallite on the wall of the pit at the top right-hand-side of Figure 1(b). The crystallite has nucleated between two atomic layers of the HOPG. (b) STM image of the HOPG surface after CVD diamond growth showing diamond nucleations on the walls of the etch pits and not on the smooth sp^2 bonded parts of the surface. The identification of the nucleations as diamond is based on Raman spectroscopy.

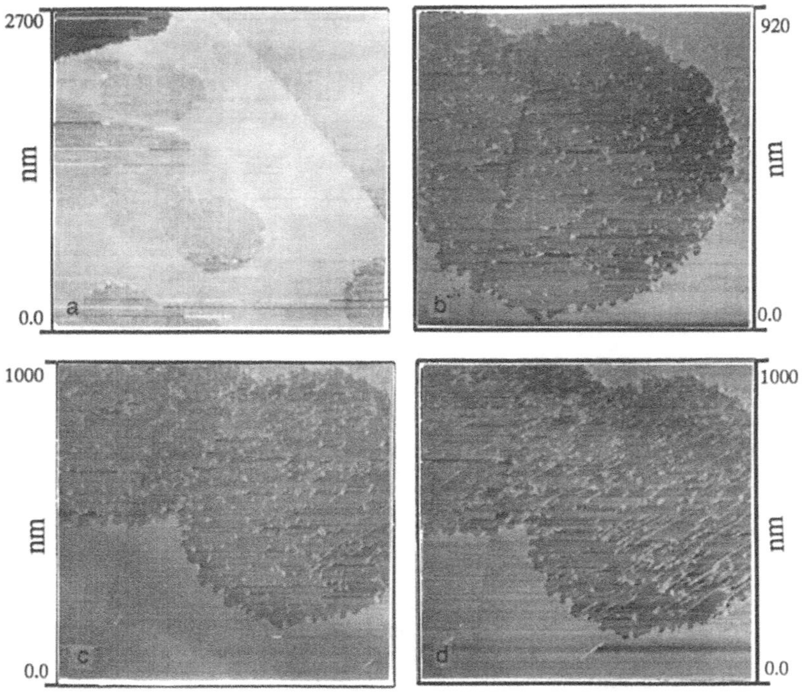

Figure 3. (a) STM image of HOPG surface after atomic hydrogen etching with the surface at room temperature. Large circular etch pits are observed. (b) Close up STM image of the central pit in (a). (c) STM image of central double-pit in (a) after biasing the tip to 10 V for 1 s. Tip position was 3.2 μ to the top right of the pit. New oriented linear nanoscale structures are observed. (d) STM image of the surface shown in (c) after biasing the tip once again to 10 V. Additional oriented linear structures are observed. We conjecture these structures are hydrocarbon chains.

Figure 1 (a) except with the HOPG at room temperature. Figures 3(a) and (b) show STM images of HOPG treated in this way. Large, nearly circular etch pits with rough edges are observed on the surface. The diameters of the etch pits are on the order of 10,000 Å, similar to the diameters of the larger pits observed in Figure 1(a). These results provide strong evidence that the shape of the etch pits is determined by the mobility of atoms on the edges of the pit At high temperatures, the atoms on the edges can move along the edges and fill in any atomic "notches" that develop on the edges as a result of the surface etching. The hexagonal shape of the pits can be explained if the directions of highest mobility have the hexagonal symmetry of the underlying substrate. At lower temperatures, the atoms on the edge do not have sufficient mobility to fill in the "notches" and therefore the etch pits grow isotropically. The rough fractal appearance of the edge, as observed in Figure 3(b), is also due to this low mobility. The mobility is not very low at room temperature, however, as indicated by the slight hexagonal shape of the pit in Figures 3(b)-(d). The small islands observed in the pit in Figure 3(b) can be explained as being formed as the edge moved outwards. When a 10 V bias was applied to the tip for 1 s, we observed the formation of oriented linear nanoscale structures in the pit, as shown in Figures 3(c) and (d). The position of the tip was approximately 3.2μ to the top right of the pit. Figure 3(c) shows an STM image of the pit in Figure 3(b) after one such tip bias. Figure 3(d) shows an STM image of the same region after an additional tip bias. The linear structures have grown in number. The direction of the structures near the bottom of Figure 3(d) makes a slightly greater angle with the horizontal than the direction of the structures near the top of the figure. The point of intersection of these directions is approximately 3.2 μ to the top right of the pit which is the approximate position of the tip. We conjecture that these linear structures are hydrocarbon chains which are oriented by the electric field of the tip. To grow diamond films on Si, the Si substrate was polished with 1μ diamond powder and cleaned with distilled water, acetone, and methanol. Diamond film growth was achieved using a filament temperature of 2000°C, sample temperature of 800°C, hydrogen gas flow rate of 300 sccm with a 0.5% methane to hydrogen mixture, and a pressure of 30 torr. STM images were acquired using tip voltages and tunneling currents of 3-4 V and 0.5-1.0 nA, respectively.

Figure 4(a) shows an STM image of a CVD diamond film grown on a Si(100) substrate for 1-1/2 hours to a thickness of approximately 2 μ. The film is polycrystalline with triangular (111) faces and square (100) faces. Scanning electron microscopy of the same film showed similar (111) and (100) faces. Figure 4(b) shows a 3-D topographic STM image of Figure 4(a) showing that the large crystallites are approximately 0.5μ in height. Using Raman spectroscopy with a laser wavelength of 5145 Å, the sample showed a sharp peak at 1332 Rcm^{-1}, shown in Figure 5(a), which is the characteristic peak for diamond. Figure 5(a) also shows a small sp^2 region at 1555 cm^{-1} indicating that some graphite is present. In comparison, Figure 5(b) shows a Raman spectrum for natural diamond with a sharp peak at 1332 Rcm^{-1}. The full-width-at-half-maximum (FWHM) for the diamond film grown on silicon and for natural diamond are 5.92 Rcm^{-1} and 3.45 Rcm^{-1}, respectively.

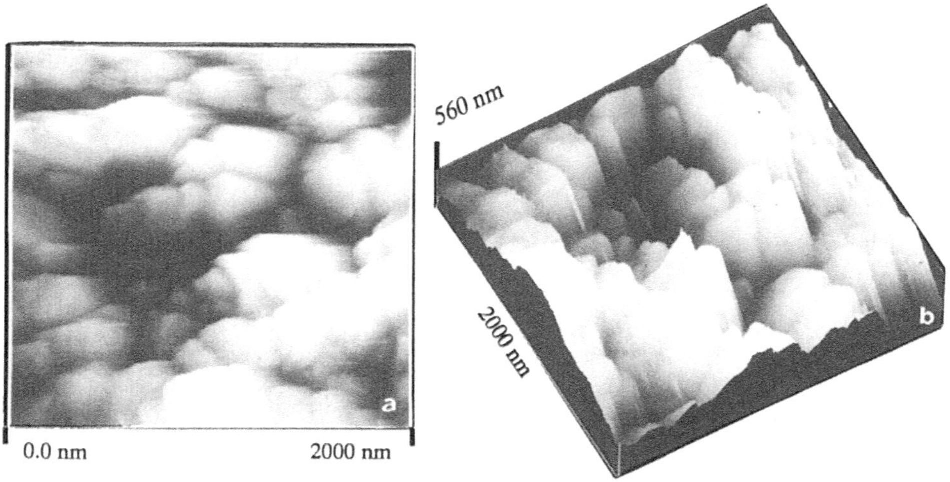

Figure 4. (a) STM image of CVD grown diamond film on Si. Diamond crystallites with triangular (111) faces and square (100) faces are observed. (b) Topographic STM image of (a). A surface roughness of approximately 0.5 μ is measured.

By gradually changing the STM tip voltage from 3-4 V to 0.4-0.5 V, we were able to acquire what we believe to be atomic resolution images of this diamond film in air. Figure 6(a) shows an STM image showing rows which are oriented at 90° with respect to one another. We believe that this is due to the 2 x 1 dimer reconstruction of the (100) surface which has been observed using LEED[6] and recently by STM in air[7] The separation distance between rows is approximately 5.12 Å, as shown in Figure 6(a). Figure 6(b) shows another atomic

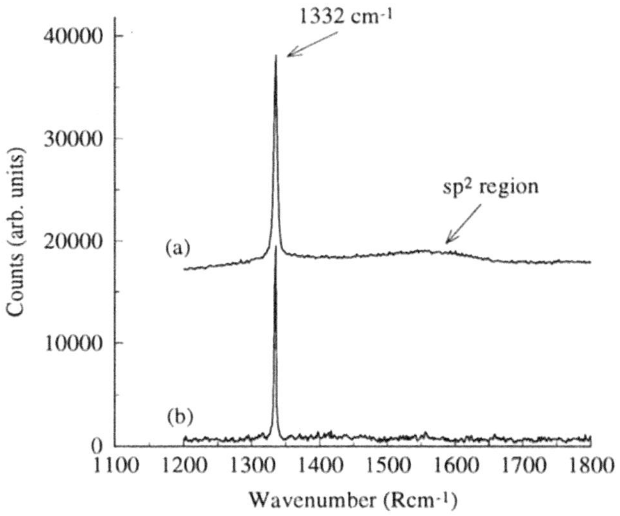

Figure 5. (a) Raman spectrum of CVD grown diamond film on Si. A sharp peak corresponding to diamond is observed at 1332 cm^{-1} with a full-width-at-half-maximum (FWHM) of 5.92 Rcm^{-1}. A smaller broader peak is observed at 1555 Rcm^{-1} corresponding to sp^2 bonded graphite. (b) Raman spectrum of natural diamond for comparison showing a diamond peak with a FWHM of 3.45 Rcm^{-1}.

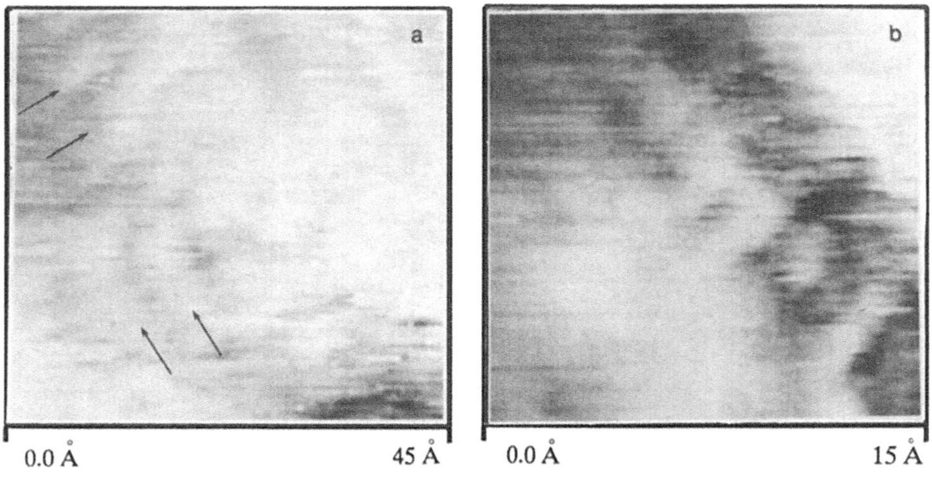

Figure 6. (a) Atomic resolution STM image of a (100) crystallite face of the CVD diamond film shown in Figure 4(a) taken in air. Parallel rows oriented at 90° are observed consistent with a 2×1 dimer reconstruction of the surface. The distance between rows is approximately 5.1 Å. (b) Another atomic resolution STM image of the (100) face showing a dimer row.

resolution image of this region. These images are consistent with sharper atomic resolution images of diamond (100) epitaxial films taken in ultrahigh vacuum (UHV).[9]

In summary, we have shown that atomic hydrogen etching of HOPG produced oriented hexagonal pits on the surface. Diamond crystallites grown using hot-tungsten filament CVD on HOPG nucleated on the walls of the etch pits and not the sp^2 regions of the surface. Large circular pits were etched on HOPG by lowering the sample temperature. We believe hydrocarbon chains formed in these pits as a result of applying a voltage to the tip. STM images of CVD grown diamond films on Si showed diamond (100) and (111) faces. Atomic resolution images of the (100) faces show a 2 x 1 dimer surface reconstruction both in air and with sharper resolution in UHV. We are currently using atomic resolution imaging of CVD grown diamond films to better understand the growth mechanism.

This work was supported in part by the National Aeronautics and Space Administration under Award No. NAG-1-1468, the National Science Foundation under Award No. DMR-9311724, and the Texas Advanced Research Program under Award No. 003594053. W. Rivera gratefully acknowledges Colciencias for financial support.

REFERENCES

1. M. Seal, in *Synthetic Diamond*, H. E. Spear and J. P. Dismukes, eds., John Wiley & Sons, New York, 507-531(1994).
2. M. N. Yoder, in *Diamond and Diamond-like Films and Coatings*, R. E. Clausing, L.L. Horton, J. C. Angus, and P. Koid, eds. Plenum Press, New York, 1-16 (1991).
3. W. A. Yarbrough and R Messier, Current issues and problems in the chemical vapor deposition of diamond, *Science*, 247:688-695 (1990).
4. N. Fujimori, T. Imai, H. Nakahata, H. Shiomi, and Y. Nishibayashi, *in Diamond, Silicon Carbide and Related Wide Bandgap Semiconductors,* J.T. Glass, R. Messier, and N. Fujimori, eds. Materials Research Society, Pittsburgh, 23-33 (1990).
5. Y. Sato, I. Yashima, H. Fujita, T. Ando, and M. Kamo, in *New Diamond Science and Technology,* R. Messier, J. T. Glass, J. E. Butler, and R. Roy, eds. Materials Research Society, Pittsburgh, 371-376 (1991).
6. Burleigh Instruments, Inc., Fishers, NY 14453.
7. P.G. Lurie and J.M. Wilson, The Diamond surface, *Surface Science,* 65:453-475 (1977).
8. T. Tsuno, T. Imai, Y. Nishibayashi, K. Hamada, and N. Fujimori, Epitaxially grown diamond (001) 2x1/1x2 surface investigated by scanning tunneling microscopy in air, *Jap. J. Appl. Phys.*, 30:1063-1066 (1991).
9. R.E. Stallcup, L.M. Villarreal, A.F. Aviles, and J.M. Perez. Atomic resolution ultrahigh vacuum scanning tunneling microscopy of diamond (100) epitaxial films, this conference proceedings.

SCANNING TUNNELING MICROSCOPY AND ATOMIC FORCE MICROSCOPY OF CHEMICAL-VAPOR-DEPOSITION DIAMOND AND DIAMOND-LIKE CARBON THIN FILMS

T. W. Mercer,[1] D. L. Carroll,[2] Yong Liang,[2]* D. Bonnell,[2]
T. A. Friedmann,[3] M. P. Siegal,[3] and N. J. DiNardo[4]**

[1]Department of Physics and Atmospheric Science
Drexel University, Philadelphia, PA 19104
[2]Department of Materials Science and Engineering
University of Pennsylvania, Philadelphia, PA 19104
**Currently:* Battelle Pacific Northwest Laboratories
POBox 999, MS K2-57, Richland, WA 99352
[3]Sandia National Laboratory, Division 1153
Albuquerque, NM 87185
[4]Department of Physics and Atmospheric Science
Drexel University, Philadelphia, PA 19104
and Department of Materials Science and Engineering
University of Pennsylvania, Philadelphia, PA 19104
***Correspondence addressee*

Abstract: Insulating polycrystalline diamond films grown by chemical vapor deposition (CVD) and semiconducting diamond-like carbon (DLC) films grown by laser-ablation have been studied using STM and AFM as well as other complementary techniques. Issues relating to the STM technique as well as the materials properties of these films have been explored. In a novel approach, photo-induced bulk carrier transport using a xenon arc lamp providing broadband radiation (λ = 180-700 nm) was used successfully to establish bulk conduction for STM imaging of the insulating films. Comparisons of topographic STM images with AFM images acquired on the same samples demonstrated the ability to correlate sub-micrometer structures observed in the images. This capability opens up the possibility that local electronic surface structure can be measured with STS. Preliminary tunneling spectra acquired on semiconducting DLC films demonstrated that illumination promotes the occupation of new electronic states resulting in a reduction of the observed energy gap at the surface. Images of DLC samples grown under a variety of conditions are providing new information into the preparation and growth parameters required to obtain the best quality films and their integrity after thermal cycling and other postfabrication treatments.

Atomic Force Microscopy/Scanning Tunneling Microscopy 2
Edited by S.H. Cohen and M.L. Lightbody, Plenum Press, New York, 1997

INTRODUCTION

Due to the unique physical, electrical and thermal properties of diamond, applications for diamond films ranging from hard, transparent coatings[1] to materials for microelectronics devices[2] are being investigated. The use of single-crystal diamond is not feasible for such applications. However, research has been directed to study means by which polycrystalline diamond and amorphous diamond-like carbon (DLC) films can be grown by different techniques such as chemical vapor deposition (CVD)[3] and laser ablation of graphite.[4,5] Bulk and surface characterization of these materials involves the determination of the sp^3 content, surface morphology, and the nature of interfaces formed with the diamond material. In this paper, we provide an overview that summarizes our on-going studies using AFM and STM directed at comparing the structural and electronic properties of surfaces and interfaces of polycrystalline diamond films grown by CVD[6] and DLC films grown by laser ablation.[5]

In order to properly relate the electronic and geometric structure of materials in which the bulk electrical conductivity is negligible, but characterization of local surface electronic structure is desired, we have employed the photoinduction of bulk carriers to demonstrate the possibility of performing STM on insulating surfaces as a first step towards eventually performing Scanning Tunneling Spectroscopy (STS) on these surfaces. We have also characterized several aspects of laser-ablated DLC films. Morphologically, these appear rather flat with reproducible surface features. We have probed a graded DLC-metal contact using STS, where we show the progression from an insulating to a metallic surface. We have also imaged a thermally-stressed laser-ablated film which de-laminated from the substrate on which it was grown. Finally, we have imaged the surfaces of the CVD and DLC films with AFM in the noncontact mode and have found a direct relationship between local electric field gradients at the surface and surface morphology.

EXPERIMENTAL

For these studies, we used both a Burleigh Aris 2200E STM and a Digital Instruments Nanoscope III equipped with STM and AFM heads. For the STM studies, we used electrochemically etched 0.5 mm diameter tungsten wire. AFM studies of surface morphology were performed in both the contact mode and the tapping mode, and electric field gradients at the surface were probed with AFM in the noncontact mode. The tips used in the AFM studies were standard gold-coated tapping mode microfabricated tips.

A hot-filament CVD process was used to grow a ~2.2 μm thick polycrystalline film on a Si substrate. The primary impurity was nitrogen which produces a deep donor level.[7] One set of laser-ablated films was grown on Si substrates with thicknesses between 100 Å and 1000 Å, and another set of laser-ablated films were grown on Pt sputter-deposited surfaces (Pt deposition on Si wafers).

PHOTO-INDUCED STM

Although AFM is the preferred technique to image the surfaces of nonconductors, STM offers the advantage to obtain local information on electronic structure. Therefore, we have investigated the feasibility of using STM to image the CVD diamond film under ultraviolet illumination. As determined by direct electrical resistance measurements, the CVD film exhibited very high resistance; illumination by broadband radiation from a 400 W xenon arc lamp, however, induced bulk electrical conduction and reduced the resistance to the point where a tunneling current can be established.

In order to achieve electron tunneling under conditions of induced bulk conduction, we focused the broadband light near the tunneling region using the geometry shown in Figure 1(a) and operated the STM. Figure 1(b) shows an unoccupied state STM image of the CVD film obtained with a sample bias of +10 V and a tunneling current of 0.6 nA acquired after thermal equilibration. Structures with lateral dimensions of ~0.1 μm and ~0.5 μm are apparent. We note that imaging was possible only under illumination, and I vs s analysis confirmed electron tunneling behavior. We have compared STM images with AFM images taken at a different place on same sample and find similar feature size features, in particular, corrugated structures ~0.1 μm in size. Measurements like these offer the possibility to perform spectroscopic measurements on insulating surfaces.

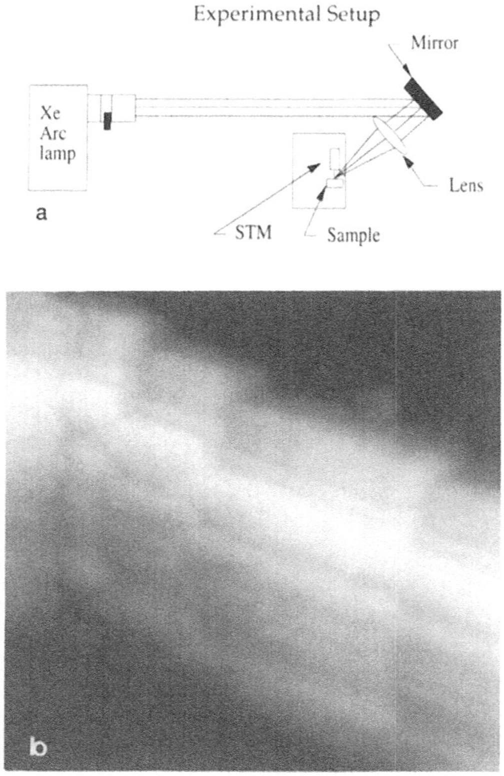

Figure 1. (a) Geometry of the experimental configuration for photoinduced scanning tunneling microscopy. (b) 0.5 x 0.5 μg^2 STM image of CVD diamond film taken at 0.6 nA demanded tunneling current with a sample bias of +10V. Light source: 400 W Xenon arc lamp with condenser housing broad spectrum between λ = 180-700 nm.

DLC THIN FILMS

This represents the first AFM studies of the Sandia-grown laser-ablated DLC films. The surfaces of the films are typically found to be very flat. For example, a 75 nm thick film exhibited features of 50-100 nm in lateral extent protruding 5 Å as shown in Figure 2. We are currently in the process of investigating the surface morphology of the DLC films of various thicknesses grown using different laser ablation parameters.

A second type of sample involved a DLC-metal interface - a DLC film grown on sputter-deposited Pt on a Si wafer. The sample was partially masked so that a graded interface between the nonconducting film and the metal substrate could be attained at the mask boundary. We found first that the DLC film adopts the morphology of the Pt film grown by sputter-deposition. In particular, Pt clusters are observed, and the DLC coating follows this morphology. Close-ups of the DLC-coated Pt clusters show subtle texture. Figure 3 shows an image across the DLC-film --> metal boundary region along with scanning tunneling spectra taken along a line across the boundary. It is clear that the insulating band gap develops gradually finally reaching ~5 eV as one progresses from the metal to the DLC film. Thermally stressing the DLC film by the xenon arc lamp resulted in delamination of the film from the substrate.

The application of an electric field above the surfaces of the CVD and DLC films and operating the AFM in a noncontact mode permitted imaging of local electric field gradients. Tapping mode images taken synchronously with the noncontact force images demonstrated highlighted structural discontinuities in the surface morphology.

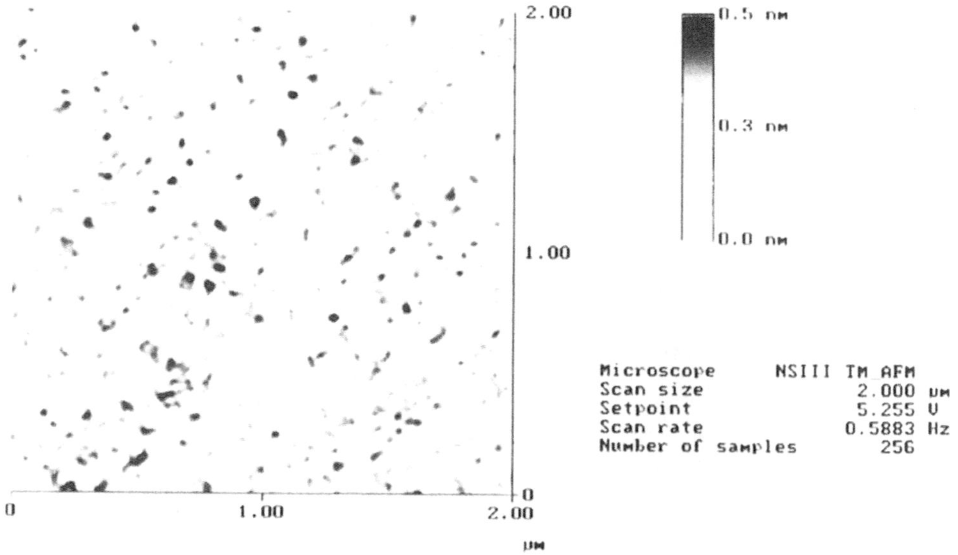

Figure 2. Laser-ablated DLC film, 75 nm thick.

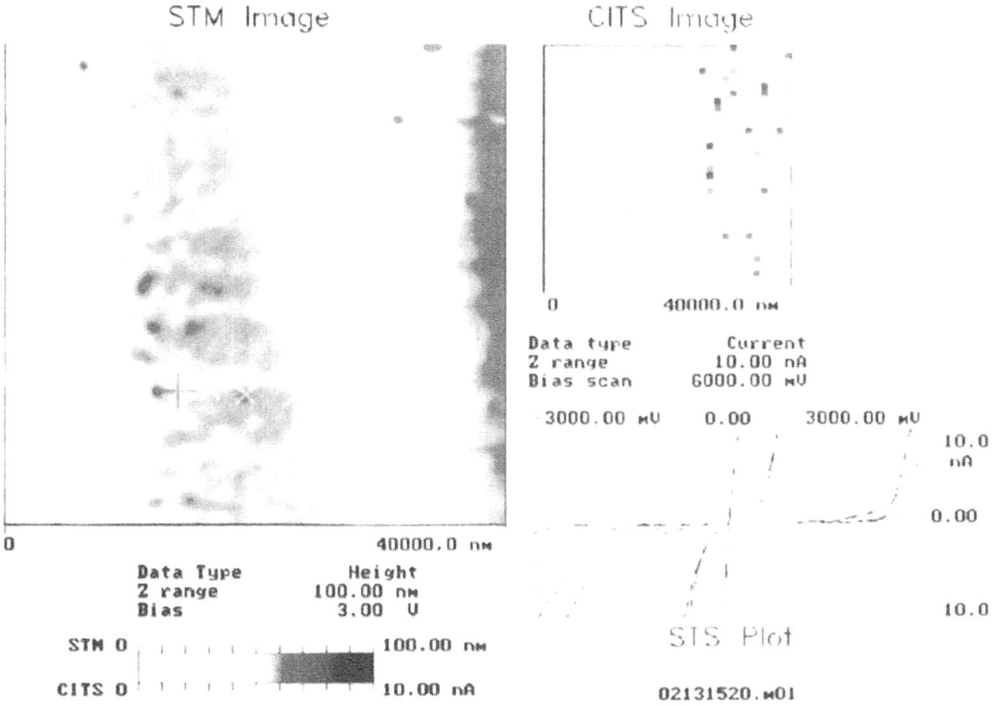

Figure 3. Image of graded DLC film grown on sputter-deposited Pt. X's denote a series of STS curves, which show a large band gap above the DLC film on the left and metallic behavior (zero band gap) of the right.

CONCLUSIONS

In this survey of studies to determine the feasibility of AFM and STM to study polycrystalline diamond films and CVD films, we have demonstrated the potential of photoinduced STM and have been able to evaluate the nanostructure --> microstructure of DLC films grown by laser ablation. Further studies will be directed at surface/interface modification and measurements of adhesion and harness. Imaging regions of negative electron affinity using STM and imaging regions of charge accumulation/depletion using noncontact AFM appear to be possible from our results.

ACKNOWLEDGMENTS

Support for this research has been provided by the National Science Foundation under grants DMR 91-20398 (NJD) and DMR 90-58557 (DAB) and from Sandia National Laboratory (NJD, TWM). The Scanning Probe Microscopy central facility at the Laboratory for Research on the Structure of Matter (University of Pennsylvania MRL) is supported under grant DMR 91-20668.

REFERENCES

1. D. C. Harris, Diamond: The ultimate durable infrared window material, *Naval Research Reviews* 44: 3-16 (1992).
2. M. N. Yoder, Diamond: It's impact on electronics, *Naval Research Reviews* 44: 17-22 (1992).
3. S. Matsumoto, Y. Sato, M. Tsutsumi, and N. Setaka, *J. Mater. Sci.* 17: 3106-3112 (1982).
4. F. Xiong, Y. Y. Wang, V. Leppert, and R. P. H. Chang, Pulsed laser deposition of amorphous diamond-like carbon films with ArF (193 nm) excimer laser, *J. Mater. Res.* 8: 2265-2272 (1993).
5. M. P. Siegal, T. A. Friedmann, S. R. Kurtz, D. R. Kurtz, D. R. Tallant, R. L. Simpson, F. Dominguez,and K.F. McCarty, Structural and electrical characterization of highly-tetrahedral-coordinated diamone-like carbon films grown by pulsed-laser deposition, Materials Research Society Spring Meeting San Diego (1994).
6. Sample obtained from Dr. Charles Beetz, ATM Corporation.
7. R. P. Messmer, and G. D. Watkins, Linear combination of atomic orbital - molecular orbital treatment of the deep defect level in a semiconductor: Nitrogen in diamond, *Phys. Rev. Lett.* 25: 656-659 (1970).

ATOMIC RESOLUTION ULTRAHIGH VACUUM SCANNING TUNNELING MICROSCOPY OF DIAMOND (100) EPITAXIAL FILMS

R.E. Stallcup, L.M. Villarreal, A.F. Aviles, and J.M. Perez

Physics Department
University of North Texas
Denton, TX 76203

Abstract: Atomic images of epitaxial (100) diamond were obtained in ultrahigh vacuum (UHV) with a scanning tunneling microscope (STM). The dimer row spacing, inner dimer pair, and single atomic step height were measured to be 0.5 nm, 0.25 nm and 0.1 nm respectively. Different forms of amorphous carbon were also observed in UHV. Some forms appeared to be randomly oriented others appeared chain-like. A radial reconstruction 1.5 nm in diameter was found on a 20° slope to a group of (100) 2x1 reconstructions. Current vs voltage spectrum was obtained in UHV and showed the electronic characteristics of the film. The Raman spectrum of the diamond film showed sp^2 and sp^3 peaks.

INTRODUCTION

A scanning tunneling microscope (STM) in ultrahigh vacuum (UHV) has many advantages over air. A surface can be cleaned in UHV and remain clean for many hours giving the STM an opportunity to get valuable and reliable atomic resolution. At normal temperatures diamond is non reactive but being a semiconductor makes it difficult to get good atomic resolution. UHV can increase the signal to noise ratio by eliminating the sample-air-tip interaction. Current vs voltage (I-V) spectroscopy is also known to be unreliable in air. The other half of the problem is the STM tip. Tips can be made of nonreactive metals such as gold or platinum but gold is too soft for good atomic resolution and platinum is extremely difficult to electrochemically etch. Platinum is also a bit too soft. Tungsten, however, is easy to etch and produces a very hard, sharp tip. The problem with tungsten is that it oxidizes in air. In UHV a tungsten tip can be cleaned and remain clean for many hours. In this paper we report, for the first time, atomic images of (100) diamond taken with an STM in UHV. We observed the 2x1 reconstruction of the (100) diamond surface. The distance between the dimer rows was measured to be 0.5 nm. We also were able to resolve the inner spacing of 0.25 nm between the dimer pairs and single atomic step heights of 0.1 nm. On the 2x1 reconstructed

Atomic Force Microscopy/Scanning Tunneling Microscopy 2
Edited by S.H. Cohen and M.L. Lightbody, Plenum Press, New York, 1997

Figure 1. UHV STM atomic resolution image of (100) diamond film. A 2x1 dimer reconstruction of the surface is observed. Dimers at 90° correspond to different single atomic steps. Also observed is an amorphous region consisting of atoms with no apparent order.

surface amorphous carbon was observed in clumps and chain-like formations. I-V spectra was also obtained in UHV with atomic resolution on a 14 nm scale. Raman spectroscopy was used to further study the nature of the diamond surface.

EXPERIMENT

A 0.25 x 1.5 x 1.5 mm (100) type 2b polished synthetic diamond substrate was purchased from Harris Corporation.[1] The substrate was cleaned[2] in acetone by ultrasonic cleaning, and also in a mixture of HCl and HNO_3 (1:3), and then in triple distilled water. Using a hot tungsten filament CVD reactor a conducting diamond film one to two micrometers was deposited on the substrate surface. The CVD reactor consists of a water cooled 4-5/8 inch five-way tee, a vertically mounted linear translator, and a horizontally mounted linear translator to shuttle the sample into the UHV STM chamber.

The reactor is coupled to the UHV chamber via an all metal through valve. Our UHV system consists of a main STM chamber pumped by a 400 L/s Varian ion pump assisted by a titanium sublimation pump with a cryoshroud. The thin diamond film was deposited for two hours with a substrate temperature of 850° C at 30 torr using hydrogen (H_2) and methane (CH_4) with flow rates of 200 standard cubic centimeters (sccm) and 1 sccm, respectively. A disappearing filament type optical pyrometer was used to determine the substrate and tungsten filament temperatures. The images were obtained in UHV (1.0×10^{-10} torr) using an Aris 5000 UHV compatible STM from Burleigh Instruments.[3] The STM probe was constructed of 20 mil tungsten wire. The tip of the wire probe was electrochemically polished in a KOH solution using the drop off technique described by Bryant.[4] Tunneling currents of 2.8 nA with a tip bias of -800 mV were needed to obtain the necessary resolution. I-V spectroscopy was obtained using the same image scan parameters (tip bias and current). For each of the five I-V spectra locations the tip was held at a constant height and the tip bias was ramped from -3V to 10V. The current was averaged

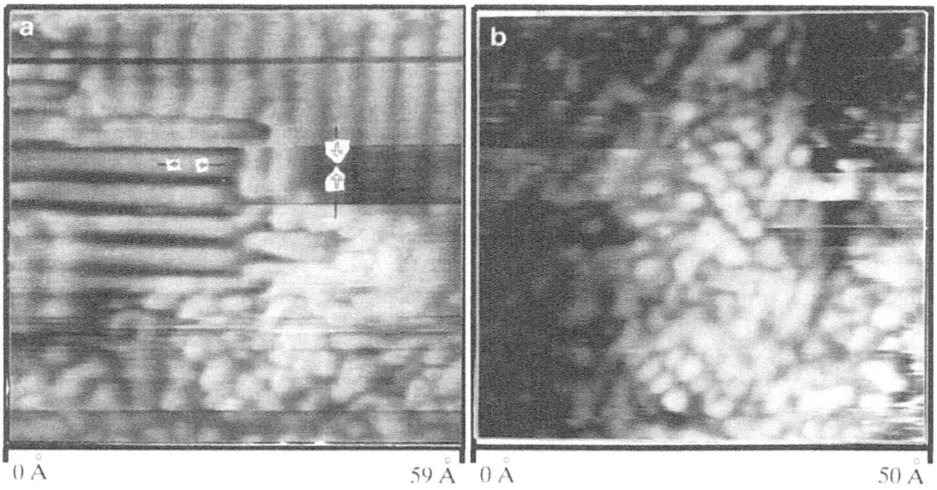

Figure 2. (a) UHV STM atomic resolution image of the same region imaged in Figure 1. Arrows indicate dimer rows where individual dimers comprising the rows can be observed. Amorphous regions are also observed. (b) UHV atomic resolution image of another region of the sample showing an amorphous region. Ordered linear chains of sp^3 bonded atoms are observed, as indicated by the arrows, in the sp^2 bonded amorphous region.

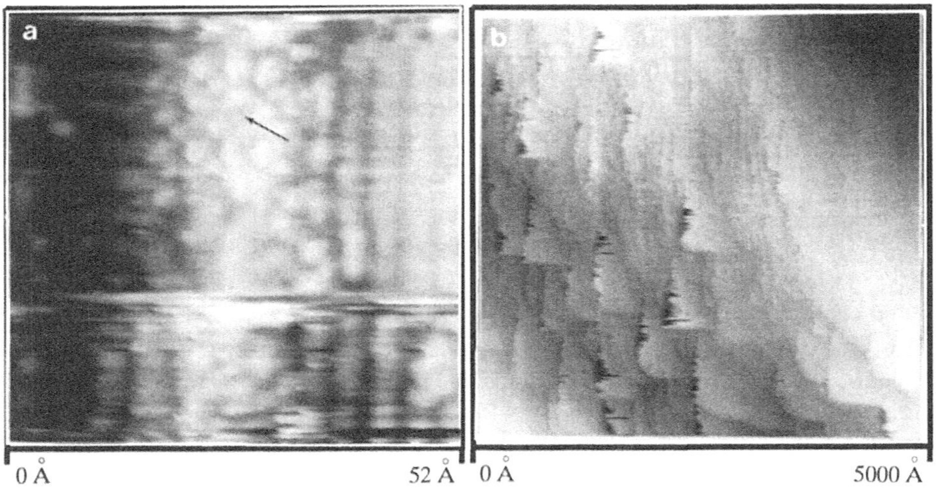

Figure 3. (a) UHV STM atomic resolution image of single molecule on the diamond (100) surface. Dimers are observed to the right and left of the molecule. (b) Large area STM image of the diamond (100) epitaxial film showing (100) planes stacked on one another and parallel to the substrate surface.

over three consecutive bias rampings for each I-V location. Raman spectra of all the specimens was carried out using a Spex 1404 double spectrometer and a Coherent Innova 90 argon ion laser.

DISCUSSION

A 2x1 reconstruction of the (100) surface consisting of dimer rows was observed as shown in Figure 1. The spacing between the rows was 0.50 nm. Reproducibility of the diamond surface on an atomic scale in UHV was easily obtained. Occasionally it was also possible to observe the inner spacing between the dimer pairs as shown in Figure 2(a) indicated by the arrows. The inner spacing between dimer pairs was measured to be 0.25 nm. This was possible only during brief periods of time. It is obvious that the tip did not change locations (i.e. tip swapping), because the image area remained consistent, only the detail changed. It may be that the density of states of the sample and tip changed momentarily. In Figure 1 steps are observed consisting of atomic planes with dimer rows

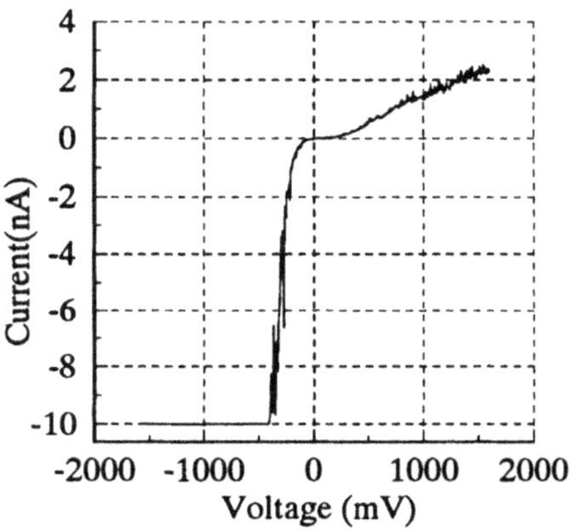

Figure 4. Tunneling current versus tip-sample voltage curve for the diamond (100) surface. The asymmetry may be due to a Schottky barrier at the surface or warping of the electronic surface states induce by the tip.

rotated in the xy-plane 90° to the dimer rows in the upper atomic plane. Step heights were 0.10 nm normal to the (100) plane (z-axis). This is consistent with STM measurements of diamond in air[5] and low energy electron diffraction (LEED).[6]

Particles grouped together in clumps with little or no apparent order were observed in the same image scan with the dimer rows as shown in Figure 1. Images of these particles were also easily reproduced. The particles appear to be similar to the amorphous carbon observed with an STM in air by Cho.[7] Figure 2(b) shows what we believe to be a disordered group of sp^2 bonded atoms with ordered linear chains of sp^3 bonded atoms. A single molecule is also observed on the (100) surface as shown in Figure 3(a), indicated by the arrow, and it is similar in shape to the reconstruction of (111) Si. One would not expect to observe a reconstruction of the (100) surface to have this geometric shape. After further analysis it is obvious that the (100) planes of Figure 3(a) are at a 20° angle to the xy-plane.

The molecule is 1.5 nm in diameter and parallel to the xy-plane. This molecule could be a reconstruction of a plane 20° to the (100) plane. Further studies will be required to determine the nature and reason for this unusual structure. The 20° inclination of the (100) planes could be due to a small localized defect on the substrate surface.

In Figure 3(b) a scan image of 0.5 x 0.5 μm shows the surface of the film to be relatively flat and epitaxial. In this figure we observe (100) planes with the same orientation stacked on one another and parallel to the substrate surface. Unlike films deposited on graphite, Si, or W[(8)] which are polycrystalline and multi-oriented, films deposited on a homogenous diamond substrate grow epitaxially. The epitaxial film is transparent. It would have been difficult to determine what side the film was grown on but the tantalum tabs used to hold the substrate during deposition served as a mask leaving small notches in the film.

Prior to the film deposition it was not possible to obtain a tunneling current from the STM tip to the cleaned diamond substrate surface. This was due to the insulating nature of undoped diamond which is typically greater than 10^{16} Ω cm.[1] Tunneling was easily achieved after the CVD diamond layer was grown. The conduction mechanism is unknown. It could be H incorporated into the lattice or some other impurity dopant, possibly from the tungsten filament. I-V spectra was obtained in UHV. Rectification of the current, as shown in Figure 4, may be due to a dopant causing a Schottky barrier[9] or warping of the electronic surface states induced by the STM probe.

Figure 5(a) shows the Raman spectrum, using a laser line of 514.5nm, of the epitaxial diamond film that our paper describes. The counts per second are represented in ln scale. This was necessary to bring out the small detail of the spectrum. The presence of amorphous carbon would explain the small observed peak of 1579 cm^{-1} corresponding to the sp^2 bond of carbon.[(10)] In comparison to the spectrum of our epitaxial film Figure 5(b) shows the Raman spectrum of polycrystalline CVD diamond film on silicon. Both films were grown using the same growth parameters. Figure 5(c) shows Raman spectrum of bulk natural diamond. Notice the absence of the sp^2 peak in natural diamond.

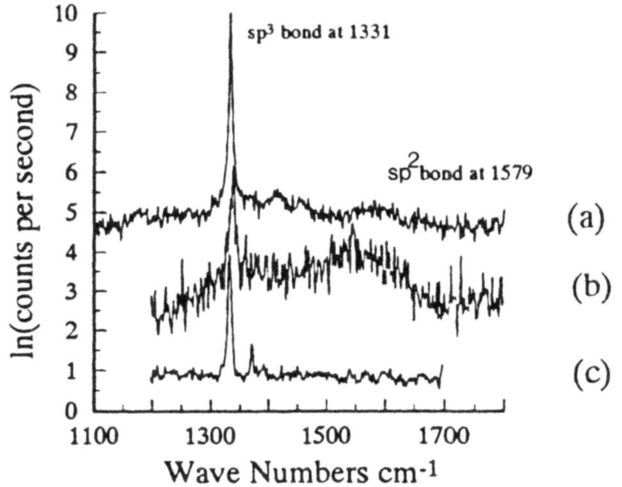

Figure 5. (a) Raman spectrum of the epitaxial diamond film that our paper describes. The small peak at 1579 Rcm^{-1} corresponds to the sp^2 bond of carbon. (b) Raman spectrum of a polycrystalline diamond film grown on Si. The peak corresponding to sp^2 bonds is larger than in (a). (c) Raman spectrum of bulk natural diamond. There is no observable sp^2 peak.

CONCLUSIONS

In summary, we have obtained the first reported atomic resolution STM images of CVD (100) diamond film in UHV. STM measurements were taken of the reconstructed (100) surface and found to be in agreement with known values. The I-V spectrum we obtained in UHV will help in understanding the electronic properties of CVD diamond film. With further study the growth mechanism of CVD diamond may be realized using the UHV STM. The carbon molecules and amorphous carbon that we observed using the STM may be a small glimpse of the complicated reaction of CVD diamond growth. Understanding the growth process will be the key in producing high quality diamond films for a wide range of applications.

This work was supported by the National Aeronautics and Space Administration under Award No. NAG-1-1468, the National Science Foundation under Award No. DMR-9311724, and the Texas Advanced Research Program under Award No. 003594053.

REFERENCES

1. Harris Corporation, 100 Stierli Court Suite 106, Mount Arlington, NJ. 07856.
2. T. Tsuno, T. Tomikawa, and S. Shikata, Diamond homoepitaxial growth on (111) substrate investigated by scanning tunneling microscope, *J. Appl. Phys.*, 75:1526-1529 (1994).
3. Burleigh Instruments, Inc., Burleigh Park, P.O. Box E, Fishers, NY 14453-0755.
4. P. J. Bryant, H. S. Kim, Y. C. Zheng, and R. Yang, Technique for shaping scanning tunneling microscope tips, *Rev. Sci. Instrum.*, 58(6):1115 (1987).
5. T. Tsuno, T. Imai, Y. Nishibayashi, K. Hamada and N. Fujimori. 1991. Epitaxially grown diamond (001) 2x1/1x2 surface investigated by scanning tunneling microscopy in air, *Jap. J. Apl. Phys.*, 30:1063-1066.
6. P.G. Lurie and J.M. Wilson, The diamond surface, *Surface Science*, 65:453-475 (1977).
7. N. H. Cho, D. K. Veirs, J. W. Ager III, M. D. Rubin, and C. B. Hopper, Effects of substrate temperature on chemical structure of amorphous carbon films, J. Appl. Phys., 71(5):2243-2248 (1992).
8. A.F. Aviles, R. E. Stallcup, W. Rivera, and J. M. Perez, Scanning tunneling microscopy of chemical vapor deposition diamond film growth on highly-oriented-pyrolytic graphite and Si, This conference proceedings.
9. J. M. Perez, C. Lin, W. Rivera, R. C. Hyer, M. Green, S. C. Sharma, D. R. Chopra and A. R. Chourasia, Scanning tunneling microscopy of the electronic structure of chemical vapor deposited diamond films, *Appl. Phys. Lett.*, 62(16):1889-1891(1993).
10. B. E. Williams, J. T. Glass, R. F. Davis, K. Kobashi and Y. Kawate, Electron Microscopy of diamond films grown by microwave PECVD, in "Extended abstracts, diamond-like materials synthesis", Johnson, Badzian and Geis, eds., Material Research Society, Pittsburgh (1988).

SCANNING FORCE MICROSCOPY CHARACTERIZATION OF BIOPOLYMER FILMS: GELATIN ON MICA

Greg Haugstad,[1] Wayne L. Gladfelter,[1] Elizabeth B. Weberg,[2]
Rolf T. Weberg,[2] and Timothy D. Weatherill[2]

[1]Center for Interfacial Engineering
University of Minnesota
Minneapolis, MN 55455
[2]E.I. du Pont de Nemours and Co.
Brevard, NC 28712

Abstract: Scanning force microscopy of thin gelatin films on mica reveals two distinct film components with characteristic frictional, morphological and stability signatures. A high-friction continuous film 1-4 nm thick strongly adheres to mica, while a low-friction component is adsorbed as porous islands on top of, or small domains within, the high-friction layer and is more easily perturbed by the scanning process. A high-force scanning procedure remarkably transforms the molecularly rough high-friction film into the molecularly smooth low-friction component if a sufficient amount of water is present in or on the film. The nanostructure of both the high- and low-friction components is imaged using a nanometer-scale asperity of gelatin attached to the SFM tip. The anticipated network structure of gelatin is observed on the high-friction layer. The low-friction material is interpreted as moieties of intramolecularly folded gelatin, with thickness (1.5±0.2 nm) equal to the diameter of the collagen-fold triple helix, containing substantial structural water.

INTRODUCTION

Scanning force microscopy (SFM) is emerging as a premier tool for characterizing the tribology,[1-3] mechanical properties[4] and surface forces[5] of organic films. Of great importance is the ability to distinguish regions differing in surface chemistry or adsorptive character via the frictional interaction with the SFM tip.[1-3] We are currently utilizing the above capabilities to characterize the structure and properties of dry and water-swollen gelatin films, and their dependence on both intrinsic and extrinsic molecular coupling mechanisms. Such information is critical to better understand and control the performance of photographic emulsions, of which gelatin comprises the binding matrix. The present work adds to our previous studies of

Atomic Force Microscopy/Scanning Tunneling Microscopy 2
Edited by S.H. Cohen and M.L. Lightbody, Plenum Press, New York, 1997

the AgBr/gelatin interface with SFM,[3,6] in turn part of recent efforts to characterize silver halide surfaces by the photographic science community.[3,6-10]

In thin gelatin films on mica we observe two film components with distinctly different frictional, morphological and stability characteristics. A high-force scanning procedure remarkably transforms the high-friction, primary component into the low-friction component. We discuss these findings based on understandings of protein folding in gelatin films.

EXPERIMENTAL DETAILS

Freshly cleaved muscovite mica (Union Mica Corp.) substrates were exposed to 10^{-3} wt% aqueous gelatin solution for three hours and dried slowly following a water rinse. The Nanoscope III (Digital Instruments) SFM was used to characterize the resulting adsorbed gelatin films. Topographic and frictional force images were simultaneously collected in air at constant vertical cantilever deflection using triangular microfabricated 100 μm cantilevers (spring constant = 0.58 N/m) with pyramidal Si_3N_4 tips. The 1231J scanner with lateral/vertical scanning ranges of 160/4.7 μm was used. Friction-actuated cantilever torsion was enabled by choosing a fast-scan direction perpendicular to the primary cantilever axis. Region-specific relative frictional forces were measured by collecting a single topography/friction trace over a left-to-right/right-to-left scanning cycle containing all surface regions to be compared. Asperity-related contributions to cantilever torsion averaged to approximately zero over the complete cycle, leaving only the nonconservative dissipative term.[3] The applied load was varied by changing the vertical cantilever deflection maintained during scanning.

RESULTS

Figure 1 contains representative topography/friction (left/right) images of as-prepared gelatin films collected at contact forces of several tens of nN. Higher elevation or frictional force is rendered brighter. Figure 1(a) surveys a 20000 x 20000 nm region revealing scattered islands exerting a reduced frictional force on the SFM tip relative to the surrounding surface. A magnified view of two islands is shown in the 3000 x 3000 nm image of Figure 1(b).

The lowest surface regions, which we will call the *first layer*, appear "granular" as imaged, the smallest resolved "grains" being several tenths of a nm in lateral dimension. A relatively large frictional force is exerted on the SFM tip in these regions. The islands contain subregions of characteristically different thicknesses. The dominant, thinner island portions have mean elevation 1.5 nm higher than the first layer; friction identical to that of the first layer is imaged at the bases of circular pores 10-100 nm in diameter. The thicker island regions have variable elevation 6-10 nm above the first layer and friction signal identical to the thinner island regions. On some samples small, low-friction domains (<100 nm diameter) were also observed *within* the first layer.[11]

Friction-load measurements quantified the relative, region-specific friction seen in Figure 1. Figure. 2 contains representative friction-versus-load data on both the first layer (open circles) and a 1.5-nm thick island (closed circles). Linear fits of first layer and island data at applied loads below ≈20 nN are shown as solid and dashed lines, respectively. Above an applied load of ≈20 nN the frictional force on the island departs from the linear trend, rising more substantially with increasing load, while the first layer data remains consistent with the initial trend.

Departure from linearity on the first layer occurs above an applied load of 100 nN. This departure from linearity generally coincided with the onset of permanent deformation (wear) observed in topography images. This led to attempts to remove film layers to quantify thickness and adhesion. The result of such a process applied to a thin strip of the film in Figure 1(b) is shown in the topography/friction images (left/right) of Figure 1(c) (also 3000 x 3000 nm), collected at a contact force of ≈25 nN. A contact force of ≈100 nN yielded the quick removal of the thinner island from the 3000-nm scanned strip, but left the first layer intact.

The result is visible near the bottom of Figure 3 (compare with Figure 1(b)). The friction and elevation of the remaining surface was identical to that probed at the bases of pores in the islands. We interpret the above process as the removal of the low-friction material to yield the underlying first layer, which apparently covers the entire mica surface.

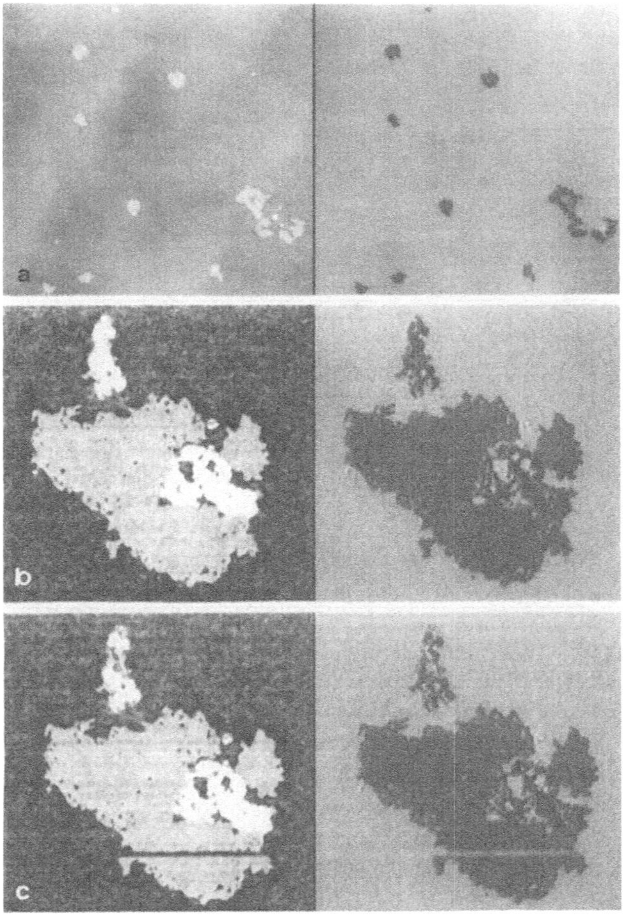

Figure 1. Representative topography/friction (left/right) images of a gelatin film; higher elevation or frictional force is rendered brighter. (a) 20,000 x 20,000 nm region displaying scattered low-friction islands; (b) a typical magnified view (3000 x 3000 nm) of two islands; (c) same as (b), following removal of selected film components via high-force scanning.

Similar results were obtained on the thicker-island regions, but at lower forces: the thick island at the top of Figure 1(b) was quickly removed from the scanned strip at a contact force of only 50 nN, leaving intact both the first layer and thinner-island portions.

At higher contact forces the first layer can be removed; Figure 3(a) (2000 x 2000 nm) shows the result of repeated raster scanning of a 1000 x 1000 nm region at ≈150 nN. The modified square region contains a discontinuous film 1.5 nm thick and is bordered by mounds of high-friction material. The mean thickness of the unperturbed film outside of this region was determined by measuring its elevation relative to the cleared mica surface (whose atomic structure was imaged). We measured first-layer thicknesses of 1-4 nm among all the films thus characterized (≈20), uniform to within ±0.5 nm for a given film. The images in Figures 3(a)-3(c) were collected approximately 1, 2.5 and 6 minutes after termination of the high-force scanning process, and record the lateral growth and coalescence of "transformed" film domains

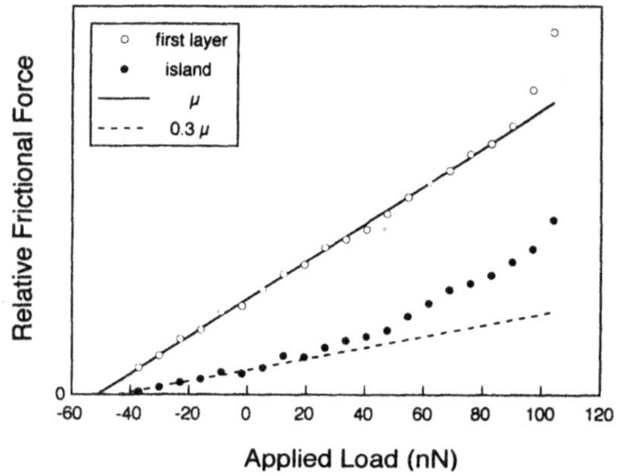

Figure 2. Frictional force (relative units) versus applied load on first-layer gelatin (open circles) and a 1.5-nm thick island (closed circles); linear fits at applied loads less than 20 nN are indicated with solid (slope=μ) and dashed (slope=0.3 μ) lines, respectively.

within the modified region. In general this evolution slowed with time and a steady-state morphology was achieved after ≈0.5-1 h. The low frictional force on the transformed film is the same as the force on low-friction islands/domains present in the as-prepared film, e.g. the 1.5-nm thick domains imaged outside of the square region in Figure 3. The mounds bordering the modified square region presumably contain removed gelatin. They display the highest friction of all surface regions and are easily perturbed by the SFM tip, accounting for changes in mound shape in Figure 3. Invoking careful procedures to prevent this, we found no correlation with the growth of the transformed films; in general the growth of the transformed films was intrinsically time dependent, i.e. not induced by SFM scanning.[11]

All such transformed films imaged (≈80), whether produced in regions 500, 1000 or 2000 nm on edge, were 1.5 ±0.2 nm thick (variance within a particular film was <0.1 nm) and exhibited friction matching the low-friction islands/domains present on the original film, irrespective of the first layer thickness (1-4 nm). We have further quantified this in measurements of frictional force versus load. Representative results at applied loads producing no wear are presented in Figure 4 for the surfaces of first layer gelatin (solid squares), a low-friction island (solid diamonds), a transformed film (open diamonds) and bare mica (solid triangles). A calibration of absolute frictional force was determined with a method described in Reference 11. A linear fit of each data set is included (solid lines). The slope of each fit yields an approximate coefficient of friction:

$\mu_{\text{1st layer}} \approx 0.5$ $\mu_{\text{island}} \approx 0.1$

$\mu_{\text{mica}} \approx 0.2$ $\mu_{\text{transformed}} \approx 0.1$

The island and transformed film values were identical within experimental uncertainty. The above ratio of $\mu_{\text{island/transformed}}$ to $\mu_{\text{1st layer}}$ (0.2) represents the lower bound of values measured on numerous samples with several SFM tips in variable humidity. The same ratio from the data in Figure 2 (0.3) is representative of the upper bound. An understanding of the contribution of tip condition, film water content, scanning speed, etc. to this variance is the subject of ongoing work.

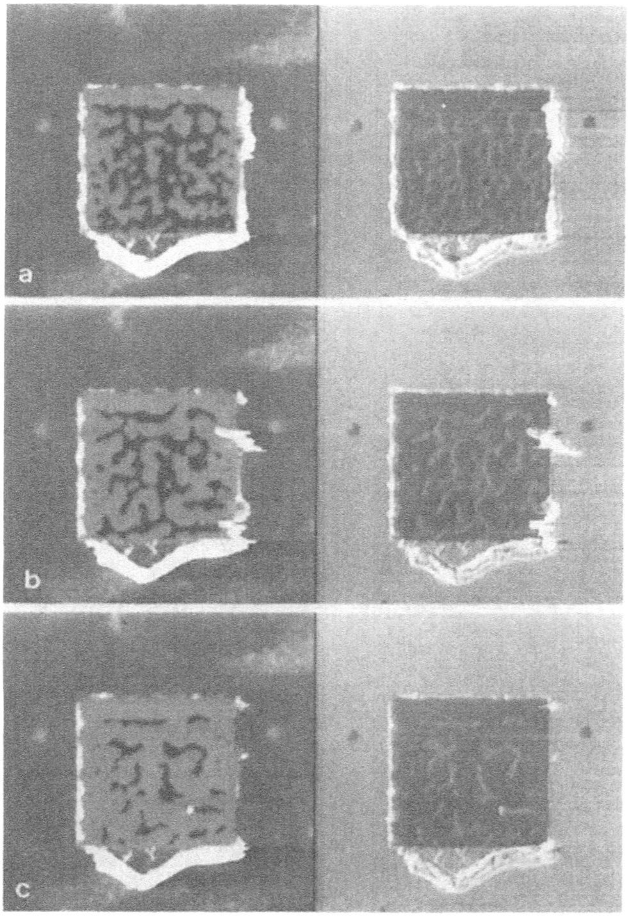

Figure 3. Topography/friction images (left/right) of a 2,000 x 2,000 nm sample region illustrating the time evolution of a subregion previously modified by high-force scanning. Elapsed times after termination of that process are approximately (a) 1 minute, (b) 2.5 minutes, and (c) 6 minutes. The resulting low-friction film is 1.5 nm thick.

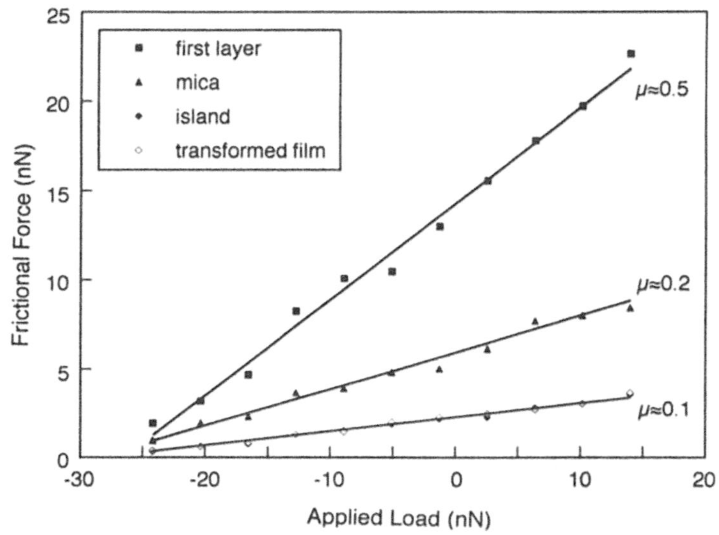

Figure 4. Frictional force versus applied load on four surface regions. A linear fit of each data set is shown (solid lines), the slope of which quantifies the frictional coefficient.

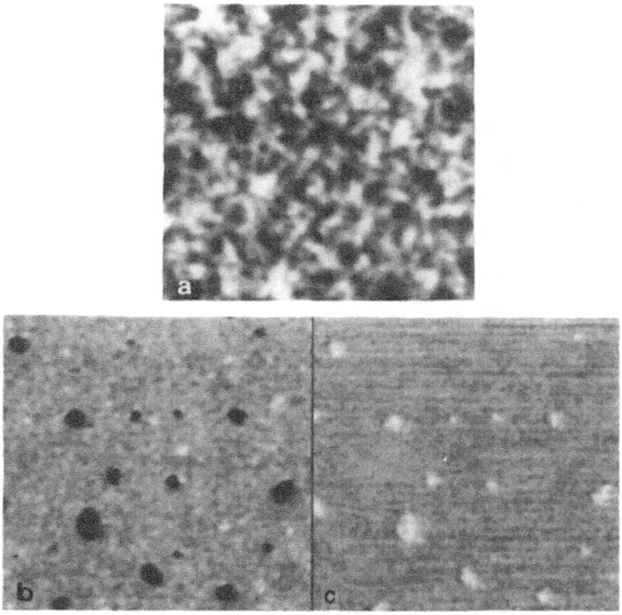

Figure 5. Representative small-scale images (300 x 300 nm) of (a) unmodified first-layer gelatin topography and (b) topography/friction (left/right) of a scan-transformed film, both obtained utilizing a nanoasperity of gelatin attached to the SFM tip.

We have successfully imaged the anticipated fibrous network structure of gelatin on the first-layer using a nanoasperity of gelatin attached to the SFM tip. (This attachment results when repeatedly ramping Z to contact forces of several hundred nN in force-displacement mode.) A representative topographic image (300 x 300 nm) is shown in Figure 5(a), with measured fiber width less than 10 nm. An amorphous network of fiber segments is revealed with typical segment length ≈20-30 nm. We also imaged the transformed films with the modified tips; Figure 5(b) contains a 300 x 300 nm topography/friction images (left/right). A faint fibrous texture is resolved on a film 1.5±0.1 nm thick containing circular pores 5-30 nm in diameter. Pores with diameters ≥10 nm display elevated friction indicating that the tip reached the mica substrate.[3]

DISCUSSION

The conventional picture of a gelatin gel is a three-dimensional network of conformationally-free polypeptide strands 0.5 nm in diameter, with rod-like triple-helical (collagen fold) crosslinks 1.5 nm in diameter.[12] The fiber width measured in SFM images is a convolution of the true fiber diameter and the shape of the imaging asperity. The molecular spacing in thin gelatin films is apparently sufficient to allow our "sharp" gelatin tips to map out individual molecular-scale fibers. The variable elevation (brightness) of the fibers in Figure 5(a) indicates three-dimensional network character even in a gelatin film only 3.3±0.5 nm thick. Gelatin networks in the thinnest films imaged (≈1 nm) displayed reduced variability in elevation, indicating more two-dimensional character.

Low-friction islands/domains observed in as-prepared gelatin films, and low-friction films resulting from the transformation of first-layer gelatin by high-force raster scanning, display many identical features, including friction coefficient, thickness, and circular pores, among others.[11] The striking similarities have led us to conclude that the materials are the same. We propose that these low-friction moieties contain intramolecularly folded, triple helical gelatin (diameter = 1.5 nm). Such folding is understood to take place in low-concentration aqueous gelatin solutions such as that employed in our film preparation (we have not observed low-friction moieties in films prepared from 1-10 wt% solutions).

The transformation from the high- to low-friction component was strongly inhibited by dry ambient conditions (RH≤30%): the low-friction film was repeatedly absent when transformations were attempted in 500 x 500 nm regions, leaving instead permanently bare mica surfaces. After exposing several samples to saturation water vapor conditions briefly, subsequent high-force scanning yielded extensive low-friction films in every attempt on each sample. Thus provided an adequate amount of water present in or on the film, intramolecular folding apparently can be induced in thin gelatin films on mica by high-force SFM scanning. The collagen-fold triple helix contains hydrogen bonding between CO and NH groups, both directly and via interstitial (structural) water molecules.[12] The net volume gain we observed for some of the transformed gelatin films suggests that some of the "free" water initially present in the vicinity but not "locked" into the film structurally (and thus not contributing to its measured thickness) is converted to structural water in the folded moiety.

In polymers like gelatin, viscoelastic behavior deriving from molecular relaxation contributes strongly to interfacial friction.[13] Gelatin molecules should have greatly reduced relaxational freedom when constrained in the triple-helical tertiary structure, whether as members of an intermolecular crosslink in a network or as intramolecularly-folded entities. We hypothesize that the folded moieties exhibit reduced friction because they contain gelatin molecules primarily, or perhaps exclusively, in the triple-helical conformation. Fluid-like character in these folded moieties[11] additionally suggests that shear deformations are not easily sustained, in which case there may be little deformation energy to be dissipated.

SUMMARY

Two distinct components in thin gelatin films adsorbed from aqueous solution onto mica were comparatively characterized:

1. A continuous "first layer" of variable thickness (1-4 nm) which exhibits a frictional coefficient more than twice as large as that on mica and which strongly adheres to mica.

2. A film component 1.5±0.2 nm thick exhibiting a frictional coefficient roughly half of that on mica and adhering to the underlying surface more weakly than the first layer; this component was manifest as
   a. porous islands up to several microns in lateral dimension located on top of the first layer,
   b. small domains less than 100 nm across located within the first layer, in contact with the mica, and
   c. films in contact with the mica after "transformation" from the high-friction first layer during a high-force raster scanning procedure.

Film nanostructure was resolved using a "nanoasperity" of gelatin attached to the SFM tip. The first layer exhibits the expected fibrous network structure of a nominally-dry gelatin gel. The low-friction component is interpreted as moieties of intramolecularly folded, triple-helical gelatin. Such protein folding apparently can be induced by high-force SFM scanning provided an adequate amount of water in/on the gelatin film.

ACKNOWLEDGMENTS

Support by the Center for Interfacial Engineering, a National Science Foundation Engineering Research Center, and a grant from E.I. du Pont de Nemours and Co., Inc., is gratefully acknowledged.

REFERENCES

1. E. Meyer, R. Overney, D. Brodbeck, L. Howald, R. Luthi, J. Frommer, H. Guntherodt, Friction and wear of Langmuir-Blodgett films observed by friction force microscopy, *Phys. Rev. Lett.*, 69:1777-1780 (1992).
2. R.M. Overney, E. Meyer, J. Frommer, D. Brodbeck, R. Lüthi, L. Howald, H. Güntherodt, M. Fujihira, H. Takano, Y. Gotoh, Friction measurements on phase-separated thin films with a modified atomic force microscope, *Nature*, 359:133-135 (1992).
3. G. Haugstad, W.L. Gladfelter, E.B. Weberg, Friction force microscopy of AgBr crystals: Ag0 rods and adsorbed gelatin films, *Langmuir*, 9:3717-3721 (1993).
4. S.A. Joyce, R.C. Thomas, J.E. Houston , T.A Michalske, R.M. Crooks, Mechanical relaxation of organic monolayer films measured by force microscopy, *Phys. Rev. Lett.*, 68:2790-2793 (1992).
5. N.A. Burnham, D.D. Dominguez, R.L. Mowery, R.J. Colton, Probing the surface forces of monolayer films with an atomic-force microscope, *Phys. Rev. Lett.*, 64:1931-1934 (1990).
6. G. Haugstad, W.L. Gladfelter, M.P. Keyes, E.B. Weberg, Atomic force microscopy of AgBr crystals and adsorbed gelatin films, *Langmuir*, 9:1594-1600 (1993).
7. H. Haefke, E. Meyer, L. Howald, U. Schwarz, G. Gerth, M. Krohn, Atomic surface and lattice structures of AgBr thin films, *Ultramicroscopy*, 42-44:290-297 (1992).
8. G. Hegenbart, T. Mussig, Atomic force microsocpy studies of atomic structures on AgBr(111) surfaces, *Surf. Sci. Lett.*, 275:L655-L661 (1992).
9. M.P. Keyes, E.C. Phillips, W.L. Gladfelter, In situ atomic force microscopy of AgBr tabular grains: surface effects of iodide, *J. Imag. Sci. Technol.*, 36:268-272 (1992).

10. H. Takada, H. Nozoye, Atomic force microscopy studies of AgBr emulsion grains, *Langmuir*, 9:3305-3309 (1993).
11. G. Haugstad, W.L. Gladfelter, E.B. Weberg, R.T. Weberg, T.D. Weatherill, Probing Biopolymers with Scanning Force Methods: Adsorption, Structure, Properties and Transformation of Gelatin on Mica, *Langmuir*, 10:4295 (1994).
12. M. Djabourov, Architecture of gelatin gels, *Contemp. Phys.*, 29:273-297 (1988).
13. L. Lee, ed., Advances in polymer friction and wear, *in*: "Polymer Science and Technology" Plenum Press, New York, vol. 5 (1974).

GASIFICATION STUDIES OF GRAPHITE SURFACE BY SCANNING TUNNELING MICROSCOPY

Deepak Tandon and Edwin J. Hippo

Department of Mechanical Engineering and Energy Processes
Southern Illinois University at Carbondale
Carbondale, Illinois 62901

Abstract: Carbon-gas reactions have been an important area of study over the past two decades. The basic mechanisms of carbon gasification are essential to understanding coal gasification. However, the studies of the inhibition and catalysis of carbon gasification have raised technical issues which have not been fully addressable until now. Some of these issues for catalytic gasification include: the effective particle size of the catalysts; the exact nature of bi-metallic catalysts; the interaction between carbon surfaces, gas, and catalyst; the percentage of the catalyst which is active; and catalyst deactivation mechanisms. The scanning tunneling microscope (STM) can address some of these issues. The purpose of this paper is to describe some preliminary work on applying STM to characterizing gasified carbon surfaces. Highly ordered pyrolytic graphite (HOPG) was gasified at 650° C for a net weight loss of 5 and 30%. The study has identified the tendency to develop high oxidation rate perpendicular to the basal plane despite the unfavorable thermodynamics of attack in this direction. The initial attack does not develop in the expected hexagonal pattern which develops after 5% burnoff. Oxygen on the surface can be identified by study of bond lengths in cross sectional analysis. Additional work is required to identify the exact nature of the oxygen structures, but the techniques developed in this study demonstrate that it is possible to follow gasification with the STM.

INTRODUCTION

Carbon-gas reactions have been an important area of study over the past two decades. The basic mechanisms of carbon gasification are essential to understanding coal gasification. The study of coal gasification is over a century old but the energy crisis of the early 1970's brought a renewed intensity to its study and development.[1] During the recent developments, catalytic gasification of coal became a prominent area of investigation[1]. The inhibition of carbon gasification became of interest with the advent of the space shuttle. A need for high strength light weight aerospace materials capable of withstanding large stresses at elevated temperatures in an oxidizing environment developed.[2]

The intense study of coal gasification provided some technical successes.The investigation of catalytic gasification of coal resulted in development of commercial processes.[3] In addition, efforts were made to increase the understanding of coal gasification.[4] Nickel was found to be a very effective methanation catalyst[5]. Alkali and alkaline earth bases were found to catalyze carbon-oxygen, steam, or carbon dioxide gasification.[1,6,-8] In general, the more highly dispersed the catalyst, the higher its activity.

E. J. Hippo and N. Murdie have previously described approaches to inhibiting gas carbon reactions.[2] Besides preventing catalysis, molecular inhibitors which occupy active sites reduce gasification rates. Ceramic coatings can also inhibit carbon gasification.

However, the studies of the inhibition and catalysis of carbon gasification have raised technical issues which have not been fully addressable until now. Some of these issues for catalytic gasification include: the effective particle size of the catalysts; the exact nature of bi-metallic catalysts; the interaction between carbon surfaces, gas, and catalyst; the percentage of the catalyst which is active; and catalyst deactivation mechanisms. Issues concerned with gasification inhibition include the fractions of sites covered with oxygen complexes, the fraction of sites covered by molecular inhibitors, and the nature of the interface between ceramic coatings and structural carbons.

Addressing these issues required a microscope capable of resolving individual atoms. The scanning tunneling microscope(STM) was invented in 1982 by Binnig and Rohrer,[9] who were awarded the Nobel Prize in physics for their work. The STM operates on the principle that a tunneling current can be established between an anatomically sharp tip and the surface electronic orbitals near the Fermi level of a sample.[10]

Depending on scan size of the STM image, resolution as low as 0.01 nm for lateral and 0.001 nm for vertical measurements can be obtained. Scan size can be varied from 14x14 μm to 2x2 nm. The depth of resolution is much greater than for SEM and TEM. In addition, little or no sample preparation is required and the instrument can be operated under ambient environments with little sample preparation. Thus, surfaces can be examined in far greater detail than previously and with much less degradation of the sample. These capabilities make STM an ideal candidate for studies of catalysis and inhibition of carbon-gas reactions.

The purpose of this paper is to describe some preliminary work on applying STM to characterizing gasified carbon surfaces. To the authors' knowledge this is the first work to obtain actual images of gasified materials. The development of these techniques is essential in learning to use the STM as a tool to study catalysis and inhibition of carbon gasification.

EXPERIMENTAL

All STM nanographs were obtained using a Nanoscope II scanning tunneling microscope manufactured by Digital Instruments of Santa Barbara CA. The specific settings used to obtain images were: a scan rate of 78.13 Hz; an integral gain of 20 mV, a proportional gain of 16 mV, and a bias voltage of 13.1 mV. The STM was operated in constant current mode. Highly ordered pyrolytic graphite, HOPG, was obtained from Union Carbide and it was cut by a diamond knife into 2x2 mm squares. The HOPG was oxidized at 650° C in a Perkin Elmer TGA-7 system until 5 and 30% burnoff was reached. An air flow rate of 20mL/min was employed.

RESULTS AND DISCUSSIONS

Of the many materials documented in STM research, the most prominent is HOPG (highly ordered pyrolytic graphite).[11,13,14, 20-31] Due to its highly ordered, layered structure, HOPG is used as a calibration material for the STM.[14,28,29] The 1.42 Å carbon-carbon bond length within the hexagonal structure of the basal planes is evident in STM images. [23] This, along with the 2.46 Å hexagon-center-to-center distance[14,28,29] are used to calibrate the instrument.

The layers of HOPG are staggered in such a way that every other atom in a basal plane is directly over another atom in the plane below it. [28] It is at these points where the basal planes are held together by weakly attractive Van der Waals forces. The theoretical spacing of the basal planes is 3.35 Å.[28] At the point where an atom sits above another atom, the top atom imaged with the STM appears as being approximately 0.6 Å lower than atoms not directly over other atoms in the plane below.

Figure 1(a) is a 2x2 nm topview basal plane image of HOPG. It shows a hexagonal pattern with a 2.46 Å distance between the bright circular points. The light red/orange and dark red areas represent the electron clouds over a carbon atom and the distance between the two carbon atoms (dark red to light red/orange) is 0.14 nm which agrees with the theoretical carbon-carbon bond length. Figure1(b) shows a surface view of HOPG basal plane. Clear hexagonal rings can be observed in this image. The center of the rings have been imaged both as the bumps and pits and it is believed that this is because of tunneling through different orbitals.

After various exposure times in air at 650° C, the size, shape, and distribution of pits and the linear dimension of the graphite were examined. The STM was used to image the pits, and to determine the amount of carbon layers gasified. Bond length measurements were used to deduce the presence of oxygen complexes at the HOPG surface.

The HOPG was studied at 5% net weight loss. In the initial stage of pit formation at 5% burnoff, the pits formed seem to align in a parallel orientation as can be seen in Figure 1(c). Only a few pits were formed over the entire 4.0 mm^2 of the sample surface. Also some pits extend three to five layers deep without enlarging.

The Horizontal distances between atoms in graphite, regardless of layer in which they are found, must be at least 1.42 Å. The ideal angle between carbon atom in the top basal plane and an adjacent atom in the next lower plane would be 67°. The angle of an adjacent atom two layers down has a 78.3 ° angle; an atom three layers down has an angle of 82°; and an atom in the fourth layer would have an 84° angle. The analyses performed on the Figure 1(c) shows that the bond lengths and angles do match for pure carbon and thus suggest the presence of oxygen. The analyses further show a pit two basal planes deep and another six basal planes deep. However the horizontal distances are not correct for carbon spacing in both the cases. The pits appear to be surrounded with oxygen complexes. This would be expected for a sample that was in a high-temperature, oxygen-containing atmosphere prior to cooling in air.

The familiar large hexagonal pits are formed by the 30% weight loss level. This is shown for the image of a 15,000x15,000 nm scan in Figure 1(d). The cross sectional analysis performed on this large pit, shows a pit diameter of nearly 9 microns and the depth of the pit is measured to be nearly 500 basal planes deep. The image also shows the steepness of the pit wall which is nearly vertical. Since oxidation in the z direction is expected to be much slower than in the x-y direction the steepness and the depth of the pit is unexpected. One possible explanation is that the pit is formed from attack on a very deep screw dislocation or grain boundary which allowed oxygen to penetrate many layers relative quickly. However scans were made of the unoxidized HOPG prior to oxidation and

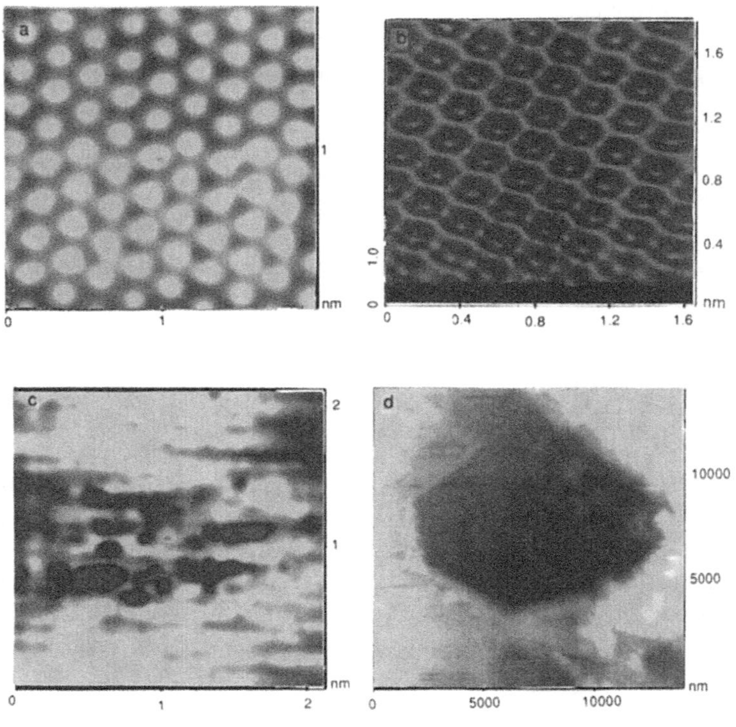

Figure 1. (a) 2x2 nm top view basal plane image of HOPG; (b) 3-D surface view of HOPG basal plane; (c)HOPG surface after 5% burnoff, showing initial stages of pit formation; (d) HOPG surface after 30% burnoff, showing large hexagonal pit

no large defects were spotted. Of course, the entire surface could not be scanned. But these pits are very common place in 30% burnoff sample.

Figure 2(a) shows a zoom inside the largest pit to a secondary pit formed inside the first pit. Additional pits can be seen inside this pit. The cross sectional analyses shows a pit diameter of nearly 8 μm. The depth of the pit nearly 300 basal planes deep. Again the image shows the steepness of the pit wall which is nearly vertical. The cross section shows deeper intrusion at the sides this might indicate that attack is occurring at a crystallite boundaries, but if true, some mechanism is needed to explain why the burnoff is proceeding faster in the z direction than the wall is receding. One would expect diffusion into the boundary to be slower than the edge recession. This is not the case. Catalytic activity might help explain this phenomena, but again why would the catalyst attack the edge slower than the grain boundary? The gasification could set up small currents which might suck oxygen to the bottom of the pit near the wall and the product gases which are known inhibitors to the gasification reactions then diffuse upward along the wall. This sets up a diffusion barrier which slows the edge recession.

Figure 2(b) is a zoom in to one of the smaller pits. The cross section profile for this pit gives an idea of the pit's diameter which is nearly 2 μm. The depth of the pit is measured to be nearly 100 basal planes deep. The image also shows the steepness of the pit wall which is nearly vertical. The cross section shows deeper intrusion at the sides. Burnoff is proceeding faster in the z direction than the wall is receding.

Figures 2(c) and 2(d) are enlarged portions of adjacent areas in Figure 2(b). These images are 4.5x4.5 nm each. From the cross sectional analysis of these images it appears that attack on pristine planes is linear. Long parallel high points are observed at the end of narrow elongated pits. These high points or streaks were originally thought to be carbon atoms, but cross-sectional analysis showed them to have oxygen complexes. The bond

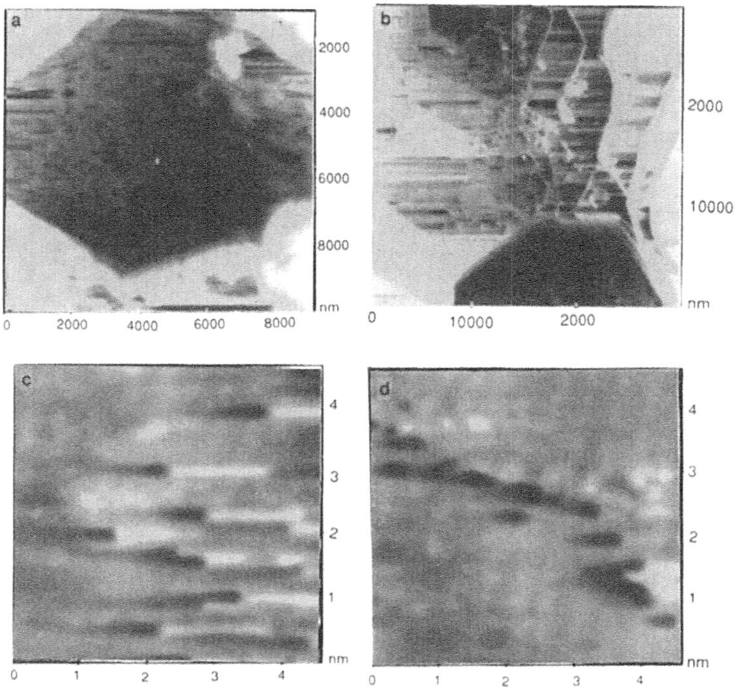

Figure 2. (a)Zoom-in of image in Figure 1(d), showing pit formation within a pit; (b) Zoom-in a small pit, showing steepness of pit well; (c)Zoom-in of image in Figure 2(b); (d) Zoom-in of image in Figure 2(b), area adjacent to one in Figure 2(c).

lengths are indicative of adsorbed oxygen molecules. The length of the streaks are multiples of 0.42 nm, which indicate attack parallel to the 101 face.

CONCLUSIONS

STM has been shown to be a powerful tool for exploring atomic structures of graphite. All six carbon atoms of the hexagonal ring structure of HOPG basal planes are imaged. Oxygen on the surface can be identified by study of bond lengths in cross sectional analysis. Additional work is required to identify the exact nature of the oxygen structures, but the techniques developed in this study demonstrate that it is possible to follow gasification with the STM. The study has identified the tendency to develop high oxidation rate perpendicular to the basal plane despite the unfavorable thermodynamics of attack in this direction. The initial attack does not develop in the expected hexagonal pattern which develops after 5% burnoff. Eddy currents may be establishing diffusional barrier slowing edge recessions of pits, but the accelerated z direction attack occurs even at the atomic scale.

ACKNOWLEDGMENTS

The Authors gratefully acknowledge the partial financial support of AMOCO. They also wish to thank Mr. Eric Sebok and Dr. Neil Murdie for their suggestions on Nanograph presentations.

REFERENCES

1. R. Meijer, "Kinetics and Mechanism of the Alkali-Catalyzed Gasification of Carbon," Ph.D., Thesis, University of Amsterdam (1992).
2. E.J. Hippo and N. Murdie, The role of active sites in the inhibition of gas-carbon reactions, *Carbon*, 27, 6, 689-695 (1989).
3. N. C. Nahas, Exxon catalytic coal gasification porcess, *Fuel*, 62, 239-241 (1983).
4. C. A. Mims, and J. K. Pabst, *Role of surface salt complexes in alkali-catalyzed carbon gasification, Fuel*, 62, 176-179, (1983).
5. P.L. Walker Jr., F. Rusinko, and L.G. Austin, The kinetics of the stereospecific polymerization of α-olefins, *Advances in Catalysis*, 11, 2-65, 1959.
6. A. Linares-Solano, E.J. Hippo, and P.L. Walker, Jr., Catalytic activity of calcium for lignite char gasification in various atmospheres, *Fuel*, 65, (6), 776-779 (1986).
7. E.J. Hippo, R.J. Jenkins, and P.L. Walker, Jr., The enhancement of lignite char reactivity to steam by cation addition, *Fuel* 57, (5), 338-344 (1979).
8. E. Hippo and P.L. Walker, Jr., Reactivity of American coal chars in carbon dioxide at 900° C *Fuel* 54, 245-248 (1975).
9. G. Binnig and H. Rohrer, Scanning tunneling microscopy, *Helv. Phys. Acta.*, 55, 726-735 (1982)
10. P. C. W. Davies, "Quantum Mechanics," Routledge and Kegan Paul, London, U.K. (1984).
11. P.K. Hansma and J. Tersoff, Scanning tunneling microscopy, *J. Appl. Phys.* 61(2), R1-R3 (1987).
12. C.J. Chen, Theory of scanning tunneling spectroscopy, *J. Vac. Sci. Technol.*. vol. A 6(2), 319-322.
13. P. Batra, N. Garcia, H. Rohrer, H. Salemink, E. Stoll, and S. Ciraci, A study of graphite surface with STM and electronic structure calculations, *Surface Science*, 181, 26-138 (1987).
14. S. Gauthier, et al. A study of graphite and intercalated graphite by STM, *J. Vac. Sci. Tech.* A 6(2), 360-362 (1988).
15. R.M. Tromp, R.J. Hamers, and J.E. Demuth, Atomic and electronic contributions to Si(111)-(7X7) STM images, *Phys. Rev.* B15, 34: 2, 1388-1391 (1986).
16. P. N. First, J.A. Stroscio, R. A. Dragoset, D. T. Pierce, and R. J. Celotta, Metallicity and gap states in tunneling to Fe clusters on FaAs(110), *Phys. Rev. Lett.*, 63: 13. 1416-1419(1989).
17. J. Schneir, H. H. Harary, J. A. Dagata, P. K. Hansma, and R. Sonnefeld, Scanning tunneling microscopy and fabrication of nanometer scale structures at the liquid-gold interface, *Scanning Microscopy*, 3:3, 719-724(1989).
18. E.B. Sebok, E.J. Hippo, N. Murdie, J.F. Byrne, and J. Rehak, Scanning tunneling microscopy of various carbon materials, Materials Technology Center 7th Annual Conference, 111-117, SIUC, April 10-11 (1991).
19. Digital Instruments, Nanoscope II Instruction Manual, Version 5, 6780 Cortona Dr., Santa Barbara, California, 93117.
20. M.R. Soto. The Effect of a vacancy on the STM image of graphite, *J. Microscopy*, 152:3, 779-788 (1988).
21. G. Binnig, et al. Energy dependent state density corrugation of a graphite surface as seen by STM, *EuroPhys. Lett.* 1(1), 31-36 (1986).
22. A. Cricient, A graphite study with a new air-operating STM, *J. Microscopy*, 152:3, 789-794 (1988).
23. V. Elings and F. Wudl, Tunneling microscopy of various carbon materials, *J. Vac. Sci. Technol.* A 6(2), 412-414 (1988).
24. L. Porte, D. Richard and P. Gallezor, STM of oxidized graphite, *J. Microscopy*, 152:2, 515-552 (1988).
25. D.P.E. Smith, H. Horber and C. Gerber, Smectic liquid crystal monolayers on graphite observed by STM, *Science*, 245, 43-45 (1989).
26. J.S. Hubacek, R.T. Brockenbrough, G. Gommie, S.L. Skala, J.W. Lyping, J.L. Latten and J.R. Shapley, STM of graphite adsorbed molecular species, *J. Microscopy*, 152: 1, 221-227 (1988).
27. P.K. Hansma, V.B. Elings and O. Marti, Scanning tunneling microscopy and atomic force microscopy:

application to biology and technology, *Science,* 242, 209-216 (1988).
28. I.P. Batra and S. Ciraci. Theoretical STM and AFM study of graphite including tip-surface interaction, *J. Vac. Sci. Techol.* A 6(2),313-318 (1988).
29. L.L. Soethout,J.W. Gerristen, P.P. McGroeneveld, B.J. Nelissen and H. van Kempen, STM measurements on graphite using correlation averaging of the data, *J. Microscopy*, 152:1, 251-258 (1988).
30. R.D. Colton, S.M. Baker, R.J. Driscoll, M.G. Youngquist, and J.D. Baldeschwieler, Imaging graphite in air by STM: role of the tip, *J. Vac. Sci. Technol.* A 6(2), 349-353 (1988).
31. S.P. Kelty, C.M. Lieber, STM investigations of the electronic structure of potassium-graphite intercalation compounds, *J. Phys. Chem.* 93: 5983-5985 (1989).

SCANNING TUNNELING MICROSCOPY STUDIES OF HYDROCARBONS ADSORBED ON GRAPHITE SURFACES

Bhawani Venkataraman and George W. Flynn

Department of Chemistry
Columbia University
New York, NY 10027

Abstract: Scanning tunneling microscopy images of n-alkanes, n-alcohols, n-alkylthiols and n-alkylchlorides have been obtained to determine if the STM is capable of differentiating between molecules where the only difference is the functional group at the ends of a molecule. These studies indicate that there are two possible ways of differentiating between these molecules: (1) by observing differences in the relative orientation of the molecules with respect to each other, and (2) by the magnitude of the tunneling current in the vicinity of the functional group. Using the tunneling current to differentiate between functional groups also permits identification of the position of the functional group in the molecule.

INTRODUCTION

Using the scanning tunneling microscope (STM) it is possible to obtain highly resolved images of individual molecules adsorbed on surfaces which reveal not only the shape of the molecules, but in certain cases their internal structure.[1,2] From these images the relative orientation of molecules with respect to each other and the surface can be determined. Information of this type has been shown to be useful in understanding both intermolecular interactions and molecule-surface interactions[1,2] since molecular orientation on a surface is governed by the interplay between these forces. Obtaining a better picture of molecule-surface and molecule-molecule interactions is important for understanding interfacial phenomena, such as chromatography, molecular epitaxy, friction, etc., which are governed by the forces between the adsorbed molecule and the surface and by intermolecular forces.

The possibility of using the STM to identify functional groups in a molecule adds a further dimension to the resolution of the instrument. For example, the possibility of performing selective chemistry becomes possible. The position of a functional group of interest can first be identified and then a reaction induced between it and a neighboring group using the tunneling electrons to excite a bond in the chosen functional group. Experiments have been performed demonstrating that the tunneling electrons can induce bond dissociation in

Atomic Force Microscopy/Scanning Tunneling Microscopy 2
Edited by S.H. Cohen and M.L. Lightbody, Plenum Press, New York, 1997

molecules.[3] Selecting a functional group for a reaction would add further specificity to the spatial resolution that is possible with the STM.

STM experiments on liquid crystal molecules containing aromatic rings suggest that the aromatic rings appear brighter than the aliphatic chains in a molecule.[4,5,6] The mechanism via which the aromatic rings appear bright in the STM images is still not completely understood. Proposed mechanisms suggest that interactions between the adsorbed molecule and the surface can change the density of states at the Fermi level of the surface.[7,8] Since the tunneling current depends on this density of states[9], an increase in this number due to the presence of an adsorbed molecule causes it to appear brighter in the STM images.

The experiments described here are aimed at investigating whether there are other functional groups that can be identified with the STM. Images of long alkyl chains terminated at one end with different functional groups (CH_3, OH, SH, Cl) were obtained to determine the effect of the functional groups on both the tunneling current and the relative orientation of the molecules with respect to each other and the underlying substrate. The molecules investigated are essentially identical (they all have a long hydrocarbon chain). The only difference between them is the functional group at the end of the hydrocarbon chain. From these images information on molecule-molecule and molecule-surface interactions can be obtained by observing variations in molecular orientation as the functional groups are varied. In addition, variations in the tunneling current in the vicinity of different functional groups can be used to distinguish between functional groups as well as to identify the position of the group in a molecule.

EXPERIMENTAL

Experiments were performed with a Digital Instruments Nanoscope III STM under ambient conditions. Triacontane ($C_{30}H_{62}$), triacontanol ($C_{30}H_{61}OH$), and 1-chlorooctadecane ($C_{18}H_{37}Cl$),were purchased from Aldrich and used without further purification. The 1-docosane thiol ($C_{22}H_{45}SH$) was synthesized by Whitesides et al.[10] Solutions of approximately 1 mg/ml in phenyloctane were prepared. Neat samples of 1-chlorooctadecane were used since it is a liquid at room temperature. A drop of solution was deposited on a piece of freshly cleaved highly ordered pyrolytic graphite (HOPG), purchased from Advanced Ceramics Corporation. The STM tips used were 0.01" diameter Pt/Rh (87/13) wire that were snipped with wire cutters. The tip was immersed in solution and the STM images obtained under a liquid drop. Images were obtained with different tips and samples to check for reproducibility and to ensure that the images were free from artifacts caused by the tip or sample.

RESULTS

Figure 1 shows an STM image of triacontane adsorbed on graphite. The image reveals the high order exhibited by these molecules on the graphite surface For the alkanes, the angle between the molecular axis and a line drawn along the troughs between two rows is 90°. This orientation as well as the intermolecular distances between molecules in the same row and adjacent rows agree with those obtained from x-ray crystallography data for solid alkanes.[11] The alkane rows run parallel to each other over areas of at least 300x300 nm^2, which is the largest area scanned in these experiments. These images agree with STM images previously published.[1,2,12]

An STM image of triacontanol molecules on graphite is shown in Figure 2(a). By looking at an individual alcohol molecule we are unable to identify at which end of the molecule the OH group is located. For the alcohols, however, the angle between the molecular axis and a line

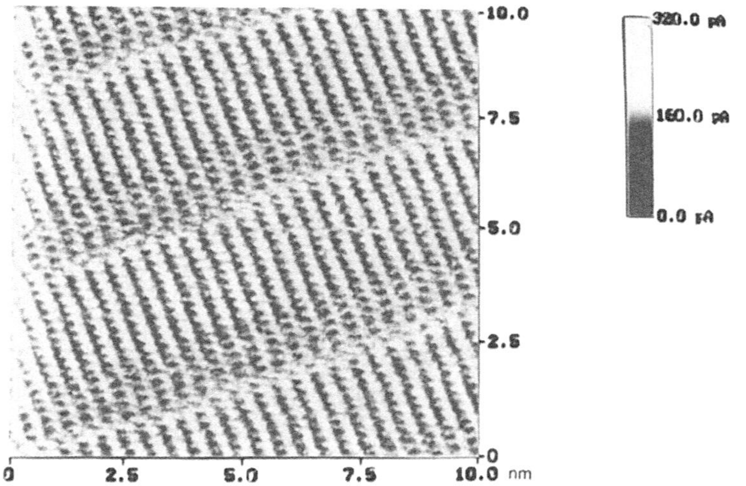

Figure 1. STM image of triacontane on graphite. The angle between the molecular axis and the troughs between two rows is 90°.

drawn through the trough between two rows is 60° as opposed to the 90° orientation observed in images of the alkanes. In addition, instead of forming straight rows (as the alkanes), the rows of alcohol molecules change their direction by 120° (Figure 2(b)) and form a "zig-zag" pattern on the graphite surface. The above observations agree with those of previously published STM images of alcohols adsorbed on graphite.[1,2,12b]

STM images of the alkylthiols on graphite are quite different from those of the alcohols. Figure 3 shows an STM image of 1-docosane thiol on graphite which reveals bright spots dispersed over the area scanned and, at lower contrast, bands connected to each bright spot. The bright spots are attributed to the SH functional group and the band connected to each

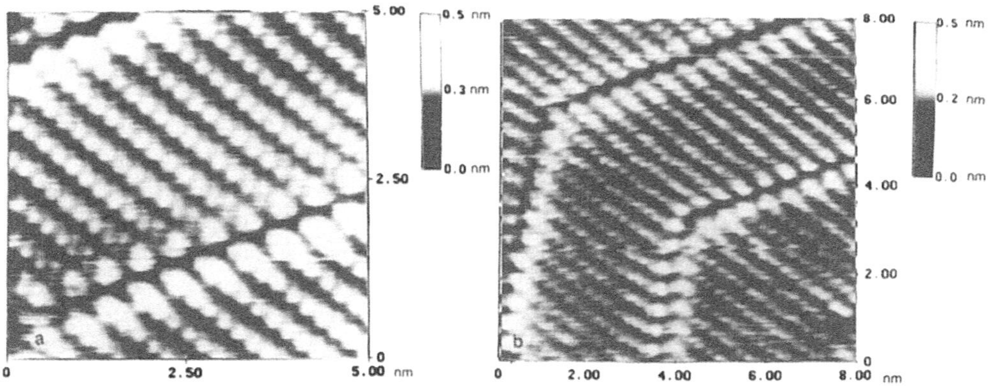

Figure 2. (a) STM image of triacontanol on graphite. The angle between the molecular axis and the troughs between two rows is 60°. (b) STM image of triacontanol on graphite showing the change in direction of the alcohol rows by 120°.

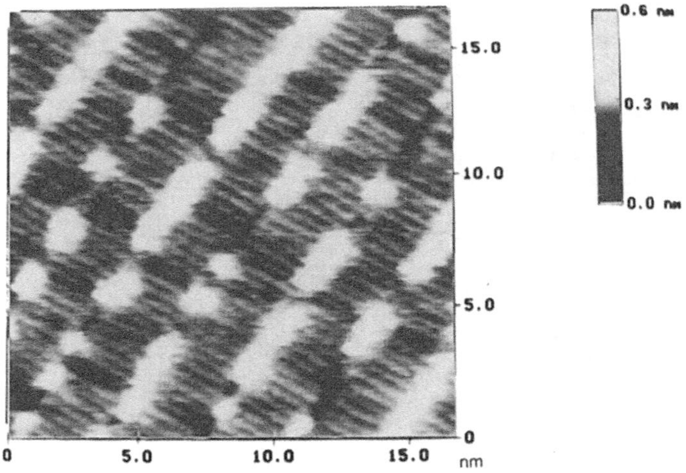

Figure 3. STM image of 1-docosane thiol on graphite. Note the bright spots dispersed over the image and at lower contrast bands connected to each bright spot. The bright spots correspond to the position of the SH group in a molecule and the band the alkyl chains attached to each SH group.

bright spot to the alkyl chains. The higher contrast observed for the SH groups is due to a larger tunneling current over the SH groups than that over the alkyl chains. This difference in tunneling current permits us to determine the position of the SH group in an alkylthiol molecule as well as the relative orientation of SH groups in adjacent molecules. From the STM images, it appears that some of the alkylthiols are arranged with SH groups in adjacent rows facing each other (i.e. head-to-head) while for others the CH_3 end of a molecule in one row faces the thiol end of a molecule in the adjacent row (head-to-tail). For the alkylthiols the angle between the molecular axis and a line drawn along the troughs between two rows is 90°, which is different from the 60° orientation exhibited by the alcohols.

STM images of the alkylchlorides appear much like those of the alkanes (Figure 4). These images do not reveal bright spots as seen in the images of the thiols and therefore the position of the Cl group cannot be determined. The angle between the molecular axis and the troughs between two rows is 90°.

DISCUSSION

The high order exhibited by these hydrocarbon molecules on the graphite surface is due to a combination of intermolecular interactions and molecule-surface interactions. These hydrocarbon molecules have unusually high heats of adsorption on graphite which has been attributed to an almost perfect match between the spacing of alternate methylene groups in the hydrocarbon chain and that of the hollows in the graphite lattice.[13] This lattice match between the molecules and the substrate helps to order the molecules on the surface. However, the orientation of molecules with respect to each other and the intermolecular distances between the molecules agree with those determined by x-ray crystallography data of these molecules.[14]

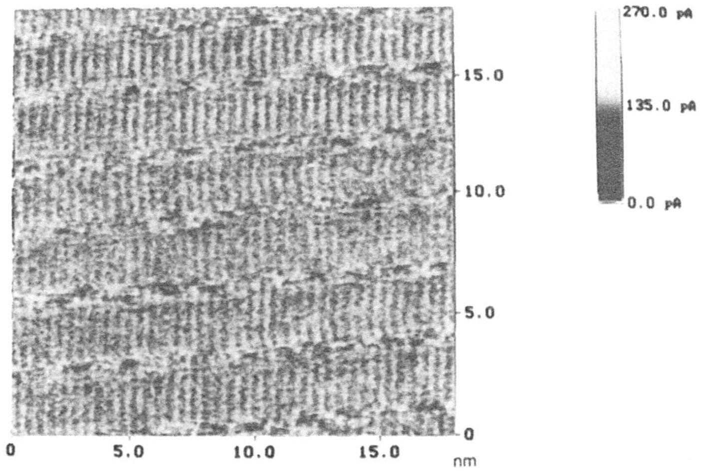

Figure 4. STM image of 1-chlorooctadecane on graphite. The angle between the molecular axis and the troughs between two rows is 90°.

This indicates that intermolecular interactions are also important in stabilizing these molecules on the surface.

Since the alkane molecules are nonpolar, the intermolecular forces are primarily van der Waals in nature. In order to optimize this interaction the molecules must lie parallel with respect to each other so as to maximize the overlap between adjacent molecules. The 90° angle between the molecular axis and the troughs between two rows is a consequence of optimizing these van der Waals interactions. In the case of the alcohols, however, the presence of the OH groups at the ends of the alkyl chains permits the molecules to form a hydrogen bond with a molecule in an adjacent row resulting in a network of hydrogen bonds formed along the alcohols rows.[1,2,14] The network of hydrogen bonds results in an additional contribution to the stabilization energy of these molecules on the surface. The 60° angle between the molecular axis and the troughs is a result of the molecular orientation necessary to optimize the hydrogen bonding interaction between alcohol molecules.[2] By orienting at 60° instead of 90° the overlap between adjacent alkyl chains of the alcohol molecules is not optimum resulting in weaker van der Waals forces between molecules. However, since hydrogen bonding is a much stronger interaction, the loss in stability due to the weaker van der Waals interaction between chains is overcome by the stabilization gained by hydrogen bonding.

The "zig-zag" pattern observed for the alcohol chains is due to a match in the 60° orientation of the alcohol molecules and the hexagonal symmetry of the underlying substrate.[1,2] Therefore, even if the rows change orientation by 120° the molecules can maintain the hydrogen bonds between each other without loss of stability. This results in two energetically equivalent orientations for the alcohol rows - one to form straight rows and the other to change direction by 120°. The alkane molecules, on the other hand, lie at 90° with respect to the troughs which optimizes the van der Waals interactions between chains. In order for the alkane molecules to form "zig-zag" patterns, as observed with the alcohols, the molecules must lie at 60°, with respect to the troughs, at the point where the rows change direction. However, a 60° orientation would reduce the van der Waals interactions between chains with nothing to be gained energetically since they cannot hydrogen bond to each other, thereby precluding the formation of "zig-zag" rows.

The alkylthiols appear to behave differently from the alcohol molecules. For the alcohols, even though it is not possible to determine the position of the OH group in the molecule, the 60° orientation suggests that the OH groups of adjacent molecules face each other in order to form a network of hydrogen bonds along the troughs. However, the STM images of the alkylthiols indicate that the SH groups of adjacent molecules need not face each other. This is evident in Figure 3 where the bright spots in the same row do not line up, and the SH group for a molecule in one row does not necessarily face that of an SH group for a molecule in the adjacent row. The fact that not all the alkylthiols align with neighboring SH groups facing each other suggests that hydrogen bonding between the SH groups does not dominate the stabilization of these molecules on the surface. In fact, it is quite likely that the thiols are not hydrogen bonded to each other. The SH group is known to be a weak proton donor with the result that either SH groups do not form hydrogen bonds or form hydrogen bonds only with strong bases (e.g. nitrogen bases like pyridine).[15]

In the case of the alkylthiols on graphite, the stabilization of the alkylthiols is then primarily due to lateral interactions between molecules in a row. This is verified by the fact that the molecular axes of the alkylthiols are at 90° with respect to the troughs indicating that the alkyl chains of molecules in the same row are parallel to each other optimizing their van der Waals interactions. This is similar to the behavior of alkanes on graphite where the orientation of molecules in a row is the same as that for the alkylthiols. Since the alkylchlorides also adopt this 90° orientation, the molecules are stabilized on the graphite surface due to van der Waals interactions.

As can be seen in the STM images in Figures 1 and 2, an alcohol and alkane molecule look essentially identical in the image; the tunneling current over the OH group is not sufficiently different from that over the CH_3 group at the other end of the molecule to determine at which end the OH functional group is located. However, the difference in molecular orientation between the alkanes and alcohols can be used as a way of differentiating the two functional groups. This has been shown to be a convenient way of identifying alcohol and alkane regions in STM images of mixtures of these molecules.[2] This suggests that one way the STM can be used to distinguish between molecules with different functional groups is by resolving the orientation of the adsorbed molecules with respect to each other. In addition, as discussed above, by resolving the relative orientation of molecules with respect to each other and the surface, detailed information on intermolecular and molecule surface interactions can be obtained. This demonstrates the power of the STM to elucidate these interactions and the interplay between them which determines molecular orientation.

While using differences in molecular orientation is one way of identifying molecules containing functional groups, it would be appealing if the position of the functional group in the molecule could also be determined. The STM images of the alkylthiols on graphite indicate that the tunneling current over the SH groups is dramatically different from that over the CH_2 groups of the rest of the molecule. This makes it possible to identify at which end of the molecule the SH group is located. Identifying the functional group also gives further insight into the intermolecular interactions between these molecules. For example, as mentioned above, the fact that not all the SH groups of molecules in the same row lie along the same line and that not all SH groups of adjacent molecules point towards each other is a result of the weaker hydrogen bonding between SH groups.

From the STM images of the alcohols and alkylthiols, it is clear that the tunneling current over the SH group is larger than that over the OH group. One obvious difference between the two groups is the higher electron density on the S atom relative to the O atom. Since the STM ultimately measures variations in electron density as the tip rasters across the surface, a possible reason for the larger tunneling current over the SH group could be the higher electron density on the S atom. If the electron density on an atom (or group of atoms) in a molecule was the only factor important in determining the tunneling current then, since the Cl atom has a larger electron density than S, it too should appear bright in the STM images. However, the STM images of the alkylchlorides do not indicate bright spots at the ends of the alkyl chains indicating that the electron density over an atom is not the only factor responsible for determining the tunneling current.

Both experimental and theoretical work indicate that the extent to which an adsorbed molecule changes the density of states at the Fermi level of the surface determines the magnitude of the tunneling current over the molecule[4,6,7,8]. The change in the density of states at the Fermi level will depend on the strength of the interaction between the molecule and the surface. Therefore, a possible explanation for the higher tunneling current over the SH groups relative to that over an OH, Cl or CH_3 group is that the SH groups interact more strongly with the graphite surface resulting in the largest perturbation of the graphite surface wave function. We are currently investigating these issues further to determine the mechanism for the increase in tunneling current when the STM tip is in the vicinity of the SH groups.

CONCLUSION

STM images of alcohols, alkanes, alkylthiols and alkylchlorides indicate that there are two possible ways of distinguishing between molecules which are essentially identical except for the functional groups at the ends of the hydrocarbon chains. The first is by resolving the relative orientation of molecules with respect to each other. Since the molecular orientations are determined by intermolecular interactions, variations in these interactions can result in different orientations. By using the STM to resolve the differences in molecular orientation as the functional group changes, it is possible to distinguish between molecules with different functional groups. The second way of distinguishing molecules with different functional groups is through the magnitude of the tunneling current over the functional group relative to the rest of the molecule. This is shown to be the case for the alkylthiols where the images reveal bright spots, which correspond to the position of the SH functional group. The mechanism via which the tunneling current over an SH group appears to be larger than that over an OH, Cl or CH_3 group may be due to a stronger interaction between the SH group and the graphite surface compared to that between the other functional groups and the graphite surface.

ACKNOWLEDGMENTS

The authors gratefully acknowledge J. Wilbur, J. Folkers and G. Whitesides for supplying the 1-docosane thiol samples. This work is supported by The Joint Services Electronics Program (U.S. Army, Navy and Air Force; DAAH 049440057), the Donors of the Petroleum Research Fund administered by the American Chemical Society, and the National Institute of Health (1 R03 RR06987-01A1). Equipment support was provided by the National Science Foundation (CHE-91-18782).

REFERENCES

1a. J.P. Rabe and S. Buchholz, Commensurability and mobility in two-dimensional molecular patterns on graphite, *Science*, 253: 424-426 (1991).

1b. S. Buchholz and JP Rabe, Molecular imaging of alkanol monolayers on graphite, *Angew. Chem. Int.Ed. Engl*, 31: 189-191 (1992).

2. B. Venkataraman, J.J. Breen, G.W. Flynn, STM studies of solvent effects on the adsorption and mobility of triacontane/triacontanol molecules adsorbed on graphite, *J. Phys. Chem.*, 99: 6608-6618 (1995).
3. G. Dujardin , R.E. Walkup, P. Avouris, Dissociation of individual molecules with electrons from the tip of a scanning tunneling microscope, *Science* 255: 1232-1253 (1992).
4. D.P.E. Smith DPE, J.K.H. Horber, G. Binnig, H. Nejoh, Structure, registry, and imaging mechanism of alkylbiphenyl molecules by tunneling microscopy, *Nature* 344: 641-644 (1990).
5. J.K. Spong, H.A. Mizes, L.J. LaComb, M.M. Dovek, J.E. Frommer, J.S. Foster, Contrast mechanism for resolving organic molecules with tunneling microscopy, *Nature* 338: 137-139 (1989).
6. H. Nejoh, Visible mechanism of liquid crystals on graphite under scanning tunneling microscopy, *App. Phys. Lett.* 57: 2907-2909 (1990).
7. A.J. Fisher, P.E. Blöchl, Adsorption and scanning tunneling microscopy imaging of benzene on graphite and MoS_2, *Phys. Rev. Lett.* 70: 3263-3266 (1993).
8. D.M. Eigler, P.S. Weiss, Schweizer, N.D.Lang, Imaging Xe with a low temperature scanning tunneling microscope, *Phys. Rev. Lett.* 66: 1189-1192 (1991).
9. J. Tersoff , D.R. Hamann, Theory of the scanning tunneling microscopy, *Phys. Rev. B* 31: 805-813 (1985).
10. The 1-docosane thiol was supplied by J. Wilbur, J. Folkers and G. Whitesides, Department of Chemistry, Harvard University, Cambridge, MA.
11. P.W. Teare, The crystal structure of orthorhombic hexatriacontane $C_{36}H_{74}$, *Acta. Cryst.* 2: 294-300 (1959).
12a. G.C. McGonigal, R.H. Bernhardt, D.J. Thomson, Imaging alkane layers at the liquid/graphite interface with the scanning tunneling microscope, *Appl. Phys. Lett.* 57: 28-30 (1990).
12b. G.C. McGonigal, R.H. Bernhardt, Y.H Yeo, D.J. Thomson D.J., Observation of highly ordered, two-dimensional n-alkane and n-alkanol structures on graphite, *J. Vac. Sci. Technol.* B9: 1107-1109 (1991).
13. A.J. Groszek, Selective adsorption at graphite/hydrocarbon interface, *Proc. Roy. Soc.* Lond. A 314: 473-498 (1970).
14. S. Abrahamsson, G. Larsson, E. von Sydow, The crystal structure of the monoclinic form of n-hexadecanol, *Acta. Cryst.* 13: 770-774 (1960).
15. G.C. Pimentel, A.L. McClellan, The Hydrogen Bond, W.H. Freeman and Company San Francisco and London: 201 (1960).

BIOLOGICAL AND CHEMICAL NANOSTRUCTURE

Moderator: C. Patrick Dunne, Sustainability Directorate
U.S. Army Soldier Systems Command,
Natick Research, Development and Engineering Center

VISUALIZATION OF THE SURFACE DEGRADATION OF BIOMEDICAL POLYMERS *IN SITU* WITH AN ATOMIC FORCE MICROSCOPE

K. M. Shakesheff,[1] M.C. Davies,[1] A. Domb,[2] C.J. Roberts,[1] A.J. Shard,[1] S.J.B. Tendler,[1] and P.M. Williams[1]

[1]Laboratory of Biophysics and Surface Analysis
Department of Pharmaceutical Sciences
The University of Nottingham, University Park
Nottingham, NG7 2RD, U.K.
[2]The Hebrew University of Jerusalem
School of Pharmacy
Jerusalem, Israel
91120

Abstract: The ability of the atomic force microscope (AFM) to observe dynamic polymer/liquid interfaces has been utilized to visualize *in situ* morphological changes occurring during the biodegradation of polymer surfaces. Using this technique we demonstrate the differential rate of degradation of amorphous and crystalline material and study the kinetics of erosion of an immiscible polymer blend. The use of atomic force microscopy in this applied area of research is furthering our knowledge of the importance of the relationship between surface morphology and degradation kinetics and promises to become an invaluable tool in the evaluation of novel biodegradable materials.

INTRODUCTION

For many biomedical and technological applications of polymers, molecular and supramolecular organization is of central importance to function. A greater understanding of this organization enables new materials with optimized properties to be designed. As a primary requirement for this design process there is a need to measure or observe polymeric organization and here, the atomic force microscope (AFM) is proving an invaluable tool. The AFM has been successfully implemented in the observation of molecular,[1-3] fibrillar,[4] phase separated [5-7] and many other polymer morphologies.[8,9] In this paper we present data showing that the AFM can be utilized in the *in situ* observation of dynamic processes at polymer surfaces.

Atomic Force Microscopy/Scanning Tunneling Microscopy 2
Edited by S.H. Cohen and M.L. Lightbody, Plenum Press, New York, 1997

The samples studied consist of novel biodegradable polymers.[10-12] These materials are employed in the formulation of advanced drug delivery systems and biomedical devices used to achieve controlled and predictable drug release in the body. In these formulations drug molecules are embedded in polymeric matrices, which are implanted or injected into the body. In the body, the biodegradable polymer undergoes chain scission at its surface due to the hydrolysis of labile bonds. This results in the erosion of the polymer surface and as the surface erodes, so embedded drug molecules become freed.

The organization of polymer molecules in these devices has a significant effect on the kinetics of erosion and hence on drug release rates.[13] For example, it has been found that crystalline regions in semicrystalline biodegradable polymers are eroded at a slower rate than amorphous regions.[14] We have used *in situ* atomic force microscopy to study this effect directly by visualizing the degradation of poly(sebacic anhydride) (PSA) in an aqueous environment. PSA is a polyanhydride with the structure shown in Figure 1. This polymer is semicrystalline and typically demonstrates a spherulitic morphology when solvent cast into thin films. Two

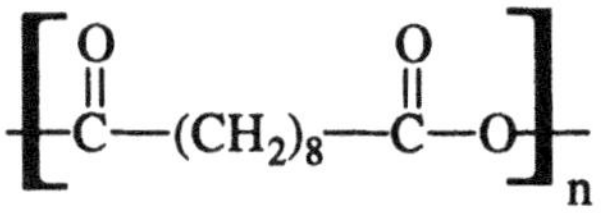

Figure 1. Chemical structure of poly(sebacic anhydride) (PSA).

experiments are described here using PSA. In the first the degradation of the pure polymer is studied with particular reference to the relative loss of amorphous and crystalline regions. The second experiment records morphological changes accompanying the degradation of an immiscible polymer blend containing equal proportions of PSA and poly(lactic acid) (PLA). The use of immiscible polymer blends in controlled release delivery systems is attracting interest[15] because it offers the potential to manipulate release kinetics simply through changes in the polymer proportions. In the PSA/PLA system, the PSA degrades at a much quicker rate than the PLA, hence by increasing the fraction of PSA in the blend it should be possible to speed up the overall degradation rate.

These results are providing a new insight into the mechanism of erosion for novel biodegradable polymers and we believe the data demonstrate the exceptional promise of *in situ* atomic force microscopy in the study of morphological changes occurring at polymer/liquid interfaces.

MATERIALS AND METHODS

Thin films of the PSA were generated by spin casting a 50 μL aliquot of a 10% (w/v) solution in chloroform onto a 1 cm x 1 cm square of freshly cleaved mica. Samples were dried in air for a period of 1 hour prior to analysis. The synthesis of the polymer has been described previously.[16]

The sample preparation was repeated in the production of the PSA/PLAN immiscible blend films by spin casting a chloroform solution containing 5% (w/v) of each of the polymers.

Atomic force microscopy was performed with the Topometrix TMX 2000 Explorer AFM (J.K. Instruments LTD., Saffron Walden, Essex, U.K.) with a 150 μm scanner head. Si_3N_4 probes on triangular cantilevers were used to obtain images in constant force mode with applied forces of not more than 1 nA and at scan frequencies of between 5 and 10 Hz.

To characterize the surface morphology of the films prior to degradation, samples were imaged in distilled water at pH 7. At this pH the degradation rate is negligible over the time period of the experiments. Degradation was performed by substituting pH 12.5 NaOH (aq) for the water. All solutions used within the imaging chamber were passed through a 0.2 μm filter before use.

RESULTS AND DISCUSSION

The morphology of spun cast PSA films consisted of tightly packed spherulites with diameters between 20 and 100 μm. The 32 μm x 32 μm AFM image in Figure 2(a) shows a cluster of these spherulites. Each one consists of a nucleating center at position A, from which bundles of fibres radiate out to fill space in two dimensions. Between the fibers and also lying over the fibers there is amorphous PSA material. This morphology has previously been recorded by scanning electron microscopy (SEM).[14]

On exposure of the film to the basic environment there was a loss of much of the amorphous material between the fibers resulting in an etching of the surface. This can be observed in Figures 2(b) and (c), in which the degradation results in the formation of elongated pits between the fibers. This result confirms observations made by SEM where the fibrous structure of the spherulites remains intact during initial degradation. However, the use of *in situ* atomic force microscopy offers a major advantage over the SEM work because for the first time the degradation can be followed on one area and in real-time. In the scanning electron microscopy studies the need to coat the sample to produce an electron conducting surface prevents morphological changes being observed on one area and therefore these changes must be inferred by the comparison of different samples.

In other experiments on PSA degradation with the AFM, we have demonstrated the pH dependence of the erosion. Also we have demonstrated that by varying the sample preparation method and so the molecular organization of the PSA it is possible to modify the erosion kinetics.

The above work on pure PSA has proven to be an excellent basis for future *in situ* AFM work because by repeating the well established pattern of erosion for spherulites we have demonstrated the suitability of the technique for the visualization of degradation. We have now moved on to analyzing novel systems such as the immiscible PSA/PLA blends where little is known about the morphological variables of degradation. The AFM images in Figure 3, demonstrate the erosion of the PSA/PLA 50%/50% thin film on exposure to the basic environment. The surface morphology of the film prior to the erosion consisted of a relatively smooth film with shallow pits. On exposure to a pH 12.5 NaOH solution, the pits deepen and broaden rapidly, until after 15 minutes the glass substrate is revealed under a polymer network. It appears that the rapidly degrading PSA is lost from the thin films resulting in exposure of a network of slowly degrading PLA. Further work using PSA/PLA at different compositions supports this theory. When a blend containing PSA/PLA 70%/30% was degraded, a similar pattern of erosion was observed; however, the remaining polymeric material after 15 minutes of degradation consisted of isolated islands, so it appears that the lower concentration of PLAN in the casting films results in the formation of PLA islands within a network of PSA.

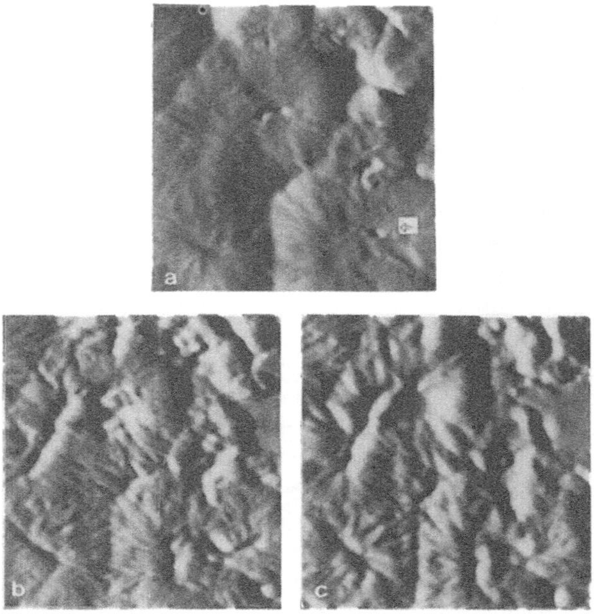

Figure 2. (a) Erosion of PSA spherulites visualized in 32 μm x32 μm AFM images taken at time 0 minutes, (b) 10 minutes, and (c) 20 minutes after exposure to a basic aqueous solution.

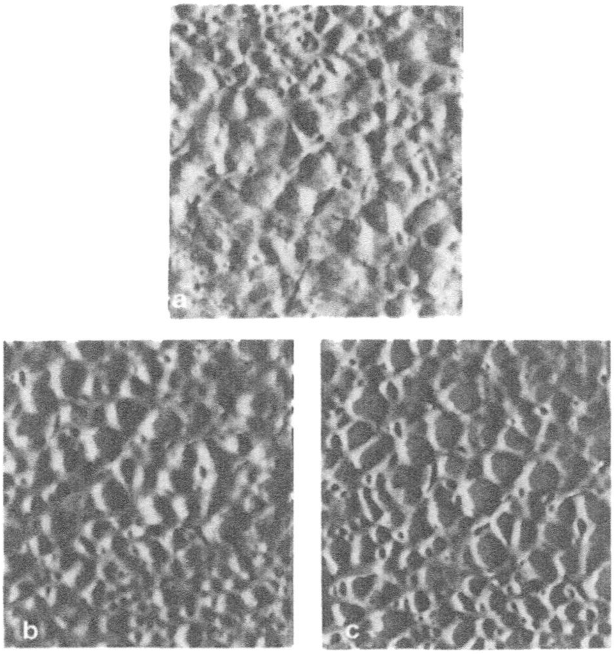

Figure 3. Immiscible polymer blend eroding. 20 μm x 20 μm AFM images taken at times (a) 0 minutes, (b) 10 minutes, (c) 20 minutes after exposure to a basic aqueous solution.

CONCLUSIONS

In situ atomic force microscopy is providing a unique insight into the morphological changes occurring at the surface of biodegradable polymeric materials. For the first time, it has been possible to visualize the loss of amorphous material from PSA spherulites on untreated samples. The technique is now generating data on novel biodegradable systems allowing the morphological factors affecting erosion to be comprehended. This work highlights the promise of the AFM in applied areas of research where an understanding of the behavior of polymer/liquid interfaces is of central importance.

ACKNOWLEDGMENTS

K.M.S would like to thank the European Research Office of the United States Army for the financial support which allowed this paper to be presented. We would like to acknowledge the support of the BRITE/Euram project, the SERC/DTI Protein Engineering LINK programme, Glaxo Group Research, VG Microtech, the DTI Nanotechnology Link programme, Kodak Limited and Oxford Molecular.

REFERENCES

1. D. Snetivy, H. Yang, B. Glomm, and G.J. Vancso, Short-range order in extended-chain crystals of polyoxymethylene from a true molecular perspective: An atomic force microscopy study, *J. Mater. Chem.* 4, 55-59 (1994).
2. B.K. Annis, J.R. Reffner, and B. Wunderlich, Atomic force microscopy of extended chain crystals of polyethylene, *J. Polym. Sci.*, Part B, 31, 93 97 (1993).

3. W. Stocker, S.N. Magonov, H.J. Cantow, J.C. Wittman and B. Lotz, Contact faces of epitaxially crystallized alpha- and gamma-phase isotactic polypropylene observed by atomic force microscopy, *Macromolecules*, 26, 5915-5923 (1993).
4. S.S. Sheiko. M. Moller, H.-J. Cantow, and S.N. Magonov, Scanning force microscopy of nanofibrillar structure of drawn polyethylene tapes: 1. Different modes and tips, *Polym. Bull.*, 31, 693-698 (1993).
5. Saraf R.F. Early-stage phase separation in polyimide precursor blends: An atomic force microscopy study, *Macromolecules*, 26, 3623-3630 (1993).
6. R.M. Overney, E. Meyer, J. Frommer. D. Brodbeck, R. Luthi, L. Howald, H.-J. Guntherodt, M. Fujihira, H. Takano, and Y. Gotoh, Friction measurements on phase-separated thin films with a modified atomic force microscope, *Nature*, 359, 133-135 (1992).
7. R.M. Overney, E. Meyer, J. Frommer. D. Brodbeck, R. Luthi, L. Howald, H.-J. Guntherodt, M. Fujihira, H. Takano, and Y. Gotoh, Force microscopy study of friction and elastic compliance of phase-separated organic thin films, *Langmuir*, 10,1281-1286 (1994).
8. B. Drake, C.B. Prater, A.L. Weisenhorn, S.A.C. Gould, T.R. Albrecht, C.F. Quate, D.S. Cannell, H.G. Hansma, and P.K. Hansma, Imaging crystals, polymers and processes in water with the atomic force microscope, *Science*, 1586-1589 (1989).
9. G.J. Leggett, M.C. Davies, D.E. Jackson, C.J. Roberts, and S.J.B. Tendler, Scanning probe microscopy of polymeric biomaterials, *Trends Polym. Sci.*, 1, 115-120 (1993).
10. K.W. Leong, and R. Langer, Polymeric controlled drug delivery, *Advanced Drug Delivery Reviews*, 1, 199-233 (1987).
11. R. Langer, New Methods of Drug Delivery, *Science*, 249, 1527-1533 (1990).
12. J. Heller, Controlled drug release from poly(ortho esters) - A surface eroding polymer, *J. Control. Release*, 2, 167-177 (1985).
13. A. Gopferich and R. Langer, The influence of microstructure and monomer properties on the erosion mechanism of a class of polyanhydrides, *J. Polym. Sci*, 31, 2445-2458 (1993).
14. E. Mathiowitz, J. Jacob, K. Pekarek, and D. Chickering III, Morphological characterization of bioerodible polymers 3. Characterization of the erosion and intact zones in polyanhydrides using scanning electron microscopy. *Macromolecules*, 26, 6756-6765 (1993).
15. K.J. Pekarek, J.S. Jacob, and E. Mathiowitz, Double-walled polymer microscopheres for controlled drug release, *Nature*, 367, 258-260 (1993).
16. A. J. Domb, S. Amselem, and M. Maniar., in *Polymeric Biomaterials,* S. Dumitriu, ed., Marcel Dekker, Inc., New York, 10016, chap. 13 (1994).

SCANNING TUNNELING MICROSCOPY INVESTIGATIONS ON HETEROEPITAXIALLY GROWN OVERLAYERS OF Cu-PHTHALOCYANINE ON Au(111) SURFACES

Torsten Fritz,[1]* Masahiko Hara,[2] Wolfgang Knoll,[2] Hiroyuki Sasabe[2]

[1]Institut fur Angewandte Photophysik
TU Dresden
Mommsenstr 13
01062 Dresden, Germany
[2]Frontier Research Program
RIKEN-Institute, Wako
Saitama 351-01, Japan
**Correspondence addressee*

Abstract: With its ultrahigh resolution, the STM provides an outstanding capability for the structural analysis of the molecular orderings occurring on single crystalline substrates.

We used the STM to image all steps of the multiple heteroepitaxial process, which leads to highly ordered and ultrathin layers of the Cu-phthalocyanine molecules on single crystalline gold films, previously grown on freshly cleaved mica sheets.

The structure of the organic overlayer is commensurable with the Au(111) surface as deduced from both reflection high-energy electron diffraction (RHEED) and STM data. The film shows a novel, nearly quadratic structure which is strongly influenced by the interaction with the Au substrate and different from that in bulky material of Cu-phthalocyanine. For the determination of the structure of organic layers the combination of *in situ* RHEED and STM is very useful and allows to overcome the limitations of both surface analysis techniques.

INTRODUCTION

For realizing currently made prospective designs of molecular devices as well as promoting basic research on low-dimensional organic solids molecular beam epitaxy (MBE) techniques in an ultra high vacuum (UHV-MBE) are a prerequisite because of the unique cleanness of the deposition process, leading to organic layers with well-defined and unique properties. Therefore, a detailed investigation of the initial growth structure of ultrathin films is essential to the understanding of the growth mechanism and heterogeneous interface formation between layers of different materials. Since the first organic molecular beam epitaxy (OMBE) was performed with *in situ* reflection high energy electron diffraction RHEED during actual film growth of copper phthalocyanine (CuPc) monolayers on MoS_2[1] considerable interest has centered on developing OMBE techniques for the controlled growth of organic assemblies at molecular length scales. Recent progress in the OMBE field has led to studies of the electronic or photonic and structural properties of hetero- or quasiepitaxially grown organic thin films deposited on several kinds of substrates as MoS_2[1,2], graphite[3-5], silicon[6], alkali halides [2,7,8] and copper[9-11], using highly stable molecules as phthalocyanines[1-4,6-11] and perylene derivates[5].

More recently, scanning tunneling microscopy (STM) has opened up an entirely new approach in the study of organic molecular systems[12,13]. The ultrahigh resolution of STM allows the direct structural analysis - even on air - of the molecular orderings occurring on single crystalline substrates. The combination of diffraction methods with real space imaging of STM is very interesting because it can reveal the real structures of epitaxially grown overlayers by overcoming the limitations of both surface analysis techniques. However, no precise discussion by means of both real and reciprocal space analyses of the heteroepitaxial growth of organic molecular thin films has been carried out so far.

Here we introduce for the first time Au(111) as a substrate for OMBE studies and report the first real images of a heteroepitaxially grown film of CuPc on gold which is commensurable with the Au(111) surface as deduced from both RHEED and STM.

EXPERIMENTAL

The heteroepitaxial growth of CuPc molecules was carried out in an UHV OMBE system, specifically designed for deposition of organic materials under a base pressure of less than 3×10^{-10} torr, as described in a previous paper[2]. This system features two relatively large chambers for separated substrate preparation and deposition of the organic film.

As a substrate for the deposition of CuPc gold was selected because prospective applications of highly ordered organic layers [14-16] (e.g., gas sensors, optoelectronic devices, organic solar cells) require metal electrodes, and it is well known that the combination CuPc-Au results in an ohmic contact[17]. Generally, single crystalline substrates are necessary to enable heteroepitaxy. Therefore, prior to the deposition of CuPc, epitaxially grown Au layers on freshly cleaved mica were prepared within the subchamber and used without breaking the vacuum. Following basic ideas of DeRose et al.[18], we obtained Au layers with flat and almost perfect (111) surfaces by the process shown in Figure 1.

After careful purification by repeated gradient sublimation (three times in a high vacuum device), CuPc was introduced into a Knudsen cell and degassed in a separate small chamber ("cell tower") at about 10^{-7} torr and increasing temperature (from 100° C to 275° C). Then the cell tower was connected to the vacuum system of the main chamber and the K-cell was lifted until a distance to the substrate of about 25 cm was reached. We kept the substrate shutter closed as the material was heated to 300° C for 5 min and then cooled down to the deposition temperature of 275° C. The deposition of CuPc itself was carried out very slowly (a few Å/h) onto the Au-substrate held at 250° C. The pressure remained below 10^{-9} torr during CuPc growth. After having cooled the sample down to room temperature we removed it from the vacuum system for further investigation by STM.

The bare Au substrate as well as the growing CuPc layer was investigated by *in situ* RHEED, using an accelerator voltage of 10 kV and an angle of incidence of a few degrees, measured to the substrate plane.

The STM system used in this study was a commercially available NanoScope II (Digital Instruments Inc., Santa Barbara, CA, USA). All images were obtained in the constant current mode in air and at room temperature, using platinum/iridium tips.

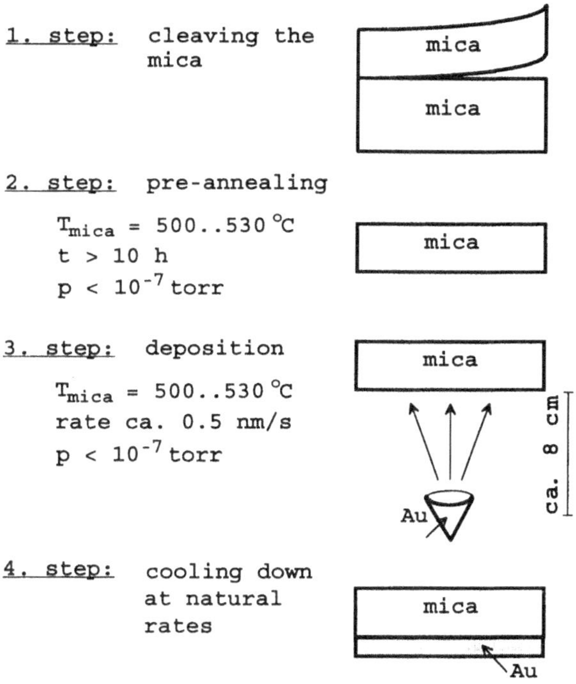

Figure 1. Preparation of epitaxially grown Au on mica.

RESULTS AND DISCUSSION

When the gold-covered mica was rotated for any multiple value of 60° in plane, the RHEED pattern remained exactly the same as compared to 0°. Since $\sqrt{3}$ times larger spacing in the diffraction pattern for the 30° direction (e.g., $[\bar{2}11]$ azimuth) and six-fold symmetry in the azimuthal rotation were observed, single crystalline growth of Au(1 11) surface with the hexagonal lattice structure could be confirmed by *in situ* RHEED.

In the case of STM imaging for such MBE grown Au(111) surface, triangular faceting where the step lines meet an angle of 60° or 120°, resulting in almost atomically flat terraces, could be observed over an area of 0.6 x 0.6 μm^2 (Figure 2(a), 2(b)). On these terraces single Au atoms could be resolved by STM occasionally (Figure 2(c)). All this indicates that a large homogeneous single crystalline area of Au(111) has been grown epitaxially on mica, thus providing very useful substrates for further heteroepitaxially grown layers.

After opening the shutter between CuPc source and substrate the intensity of the Au(111) streaks in the RHEED pattern from the $[\bar{1}10]$ azimuth decreased gradually, while a new pattern appeared between the Au(111) streaks. If the surface of the Au substrate is just covered by approximately one monolayer of CuPc molecules very sharp and narrow streaks are obtained, originating from a molecular flat CuPc layer. The distance between neighboring narrow streaks was nearly equal to 1/6 the Au streak spacing in this lattice direction and could therefore be calculated as $6 \times \sqrt{3}\ a_{Au}/2 = 14.9$ Å. When the substrate was rotated by 60° in the plane, the RHEED pattern was exactly the same as that at 0° for both Au(111) and CuPc. For the 30° direction almost the same streak distance as in the case of 0° incidence was obtained for CuPc. In any case, such angular dependencies of RHEED patterns suggest that the CuPc layer is a triple-directional crystalline molecular film with nearly a quadratic lattice whose

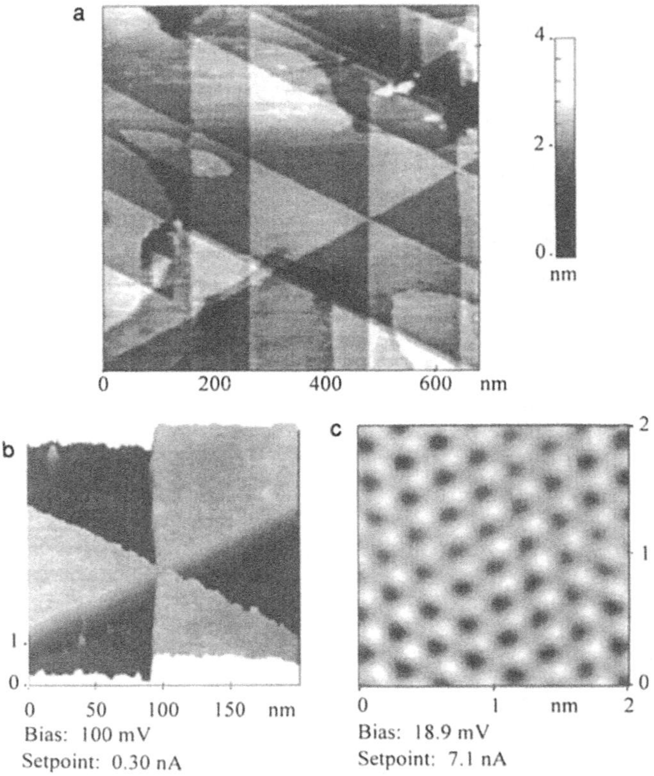

Figure 2. Epitaxially grown Au on mica (a) large area; (b) enlarged detail; (c) single Au atoms.

dominant **α**-axes are parallel to [$\bar{1}$10], [$\bar{1}$01] and [0$\bar{1}$1], respectively, reflecting the hexagonal structure of the Au(111) substrate. However, from those *in situ* RHEED observations clear evidence of both positional and orientational short-range order of the molecules can not be deduced, because organic molecules can not provide sufficient diffraction efficiencies due to their large lattice constants, resulting in quite narrow and rather faint streaks compared to the RHEED patterns obtained from inorganic crystals.

In order to determine the actual lattice structure of CuPc on Au(111) surfaces, STM imaging was carried out in air, right after finishing the deposition and cooling down the sample. Figure 4 shows two samples of STM images of CuPc overlayers. Individually distinguishable patterns with fourfold symmetry in a regular alignment are observed. The diameter of each pattern is about 13 Å, which agrees well with the known molecular structure of CuPc (Figure 3). While there exist some discussions concerning the STM imaging mechanism for organic adsorbates, i.e., the reliability of the interpretation of the images, we are sure that each four-leaf-like pattern in this study presents an individual CuPc molecule.

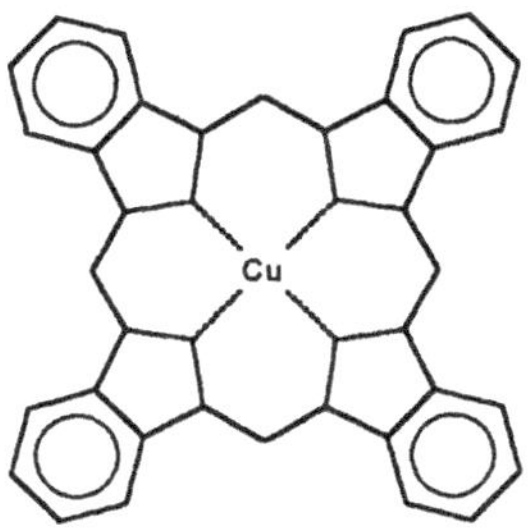

Figure 3. Molecular structure of CuPc.

According to Lippel,[11] the Cu ion can not be seen in the STM images, i.e., there is no contrast between the Au substrate and the Cu ion.

Very interesting is the finding of "negative" contrast in some STM images: the molecules are dark and the Au substrate appears bright. At the first glance, this is surprising because the bias voltage wasn't changed between the two images shown in Figure 4. If we follow the local theory of STM imaging of Spong et al.[13], we can explain this finding by assuming a temporary covering of the tip by few CuPc molecules, which will result in a local modification of the work function of the tip and inverse the contrast. The temporary covering of the tip may be caused by the decreased distance of the tip to the surface for the increased setpoint current in Figure 4(b).

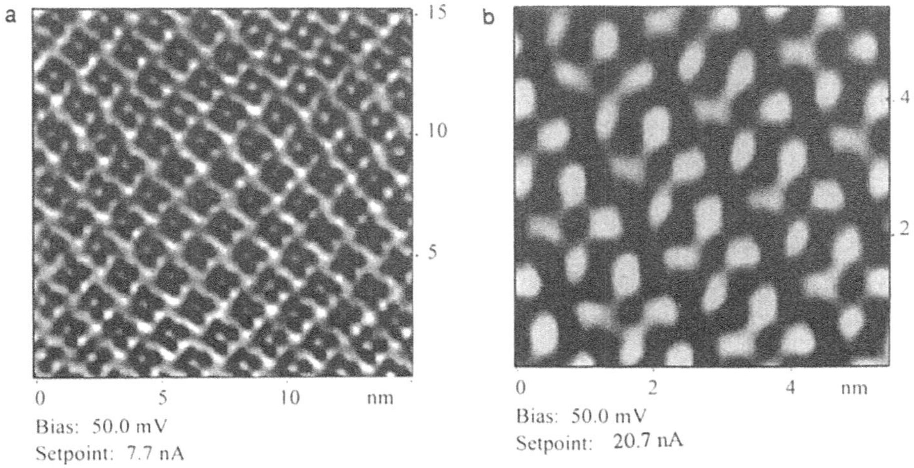

Figure 4. STM images of a CuPc monolayer on Au(111).

The STM images of the OMBE CuPc layer on Au(1$\bar{1}\bar{1}$) reveal a periodic and nearly quadratic structure with a lattice angle of 90±5° and a lattice constant of 14.5±1.5 Å. Furthermore, directions of all four-leaf-like parts are ordered in a parallel fashion to each other and inclined at an angle of approximately 30° with respect to dominant α-axes. While nonlinearity, angle distortion and drift of the STM head should be taken into account, Figure 5 shows the most reliable model structure deduced from STM images following several calibrations against Au, MoS_2 and graphite substrates with atomic resolution and *in situ* RHEED observation.

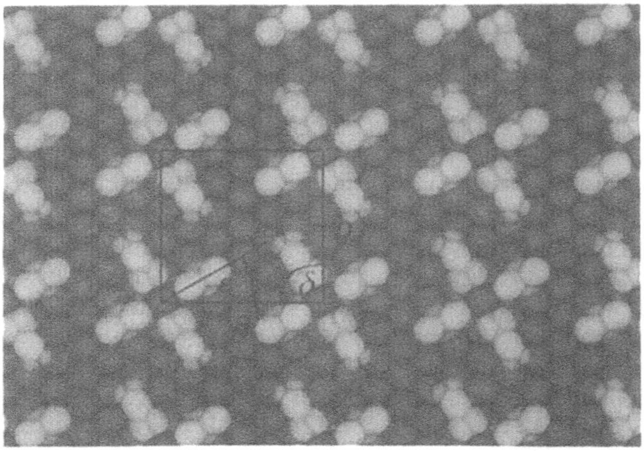

Figure 5. Model for the heteroepitaxy of CuPc on Au(111) surfaces: a = $6 \times \sqrt{3}/2 \times a_{Au}$ = 14.96 Å; b = $5 \times a_{Au}$ = 14.40 Å; δ = 90°, ε = 30°

It should be noted that our experimental results do not reveal any information about the actual place of a CuPc molecule on the Au substrate. Therefore, we assume that the Cu atom lies on top of a gold atom without any strong reason, but, in spite to other possibilities, this alignment leads to a completely symmetrical situation for opposite arms of the molecule and can furthermore be used to explain the appearingly different lengths of the molecule arms in our STM images: for the two pairs of arms the situation with regard to the underlying substrate is slightly different, as can be seen from our simple model calculation in Figure 5.

The experimentally found periodic structure comprises a large homogeneous domain over dozens of repeating units, exceeding the limited area with molecular resolution: the OMBE CuPc overlayer forms a homogeneous and continuous film covering the entire deposition area on Au(111), having surfaces smooth beyond comparison with those of normally vacuum evaporated organic thin films.

The important findings in this study are not only the high resolution molecular images but also the commensurable structure of the CuPc layer with the Au(111) surface, resulting in "artificial" lattice structures of organic ultrathin films being different from those in bulk, i.e., the obtained lattice structure is different from the structure of the well-known α-CuPc and β-CuPc crystals[19]. This is one of the reasons why experiments to obtain thicker well-ordered CuPc layers, i.e., bulky heteroepitaxially grown films, have not been successful so far.

CONCLUSIONS

Epitaxially grown Au layers on mica exhibit large single crystalline domains with flat Au(111) surfaces and are very suitable substrates for OMBE.

For the first time the heteroepitaxial growth of a CuPc overlayer on Au(111) could be proved directly by STM imaging with submolecular resolution. The overlayer shows a novel, nearly quadratic structure which is strongly influenced by the interaction with the Au substrate and different from that in bulky materials. For the determination of the structure of organic layers the combination of *in situ* RHEED and STM is very useful.

ACKNOWLEDGMENTS

The support of the Science and Technology Agency of Japan, which provided a STA-fellowship to the first author, is gratefully acknowledged. The research project was partially granted by the *Bundesministerium fur Wissenschaft und Technologie* (project No. 0329288A).

REFERENCES

1. M. Hara, H. Sasabe, A. Yamada, and A.F. Garito, Epitaxial growth of organic thin films by organic molecular beam epitaxy, *Jap. J. Appl. Phys.*, 28:L306-L308 (1989).
2. M. Hara, H. Sasabe, A. Yamada, and A.F. Garito. Observation of heteroepitaxially grown organic ultrathin layers on inorganic substrates by in situ RHEED and by STM, *Mat. Res. Soc. Symp. Proc.*, 159:57-62 (1990).
3. H. Saijo, T. Kobayashi, and N. Uyeda. Epitaxial growth of a new polymorph of Cu phthalocyanine on graphite, *J. Crys. Growth*, 40:118-124 (1977).
4. W. Mizutani, Y. Sakakibara, M. Ono, S. Tanishima, K. Ohno, and N. Toshima, Voltage dependent STM, *Jap. J. Appl. Phys.*, 28:L1460-L1463 (1989).
5. C. Ludwig, B. Gompf, W. Glatz, J. Petersen, W. Eisenmenger, M. Möbus, U. Zimmermann, and N. Karl, Video-STM, LEED and X-ray diffraction investigations of PTCDA on graphite, *Z. Phys. B.-Condensed Matter*, 88:397-404 (1992).
6. H. Sasabe, T. Wada, M. Hara, T. Furuno, Y. Aoyagi, and A. Yamada, Nanostructure fabrication of organic substances by MBE, LBF and ion beam techniques, *in:* Molecular Electronic Devices, P.L. Carter, ed., Elsevier, Amsterdam, 57-68 (1988).
7. H. Tada, K. Saiki, and A. Koma, Preparation and characterization of Vanadyl-phthalocyanine ultrathin films grown on KBr and KCl by molecular beam epitaxy, *Jap. J. Appl. Phys.*, 30:L306-L308 (1991).
8. M. Möbus, N. Karl, and T. Kobayashi, Structure of perylene-tetracarboxylic dianhydride thin films on alkali halide crystal substrates. *J. Crys. Growth*, 116:495-504 (1992).
9. J. C. Buchholz and G. A. Somorjai, The surface structures of phthalocyanine monolayers and vapor-grown films, A low-energy electron diffraction study, *J. Chem. Phys.*, 66:573-580 (1977).
10. T. Kobayashi, Y. Fujiyoshi, and N. Uyeda, The observation of molecular orientations in crystal defects and the growth mechanism of thin phthalocyanine films, *Acta Cryst.*, A38:356-362 (1977).
11. P.H. Lippel, R.J. Wilson, M.D. Miller, Ch. Wall, and S. Chiang, High-resolution imaging of copper-phthalocyanine by Scanning-Tunneling Microscopy. *Phys. Rev. Lett.*, 62:171-174 (1989).
12. J.E. Frommer, STM and AFM: applications in organic chemistry, *Angewandte Chemie, intern. Edit.*, 31:1289-1328 (1992).
13. J.K. Spong, H.A. Mizes, L.J. LaComb, M.M. Dovek, J.E. Frommer, and J.S. Foster, Contrast mechanism for organic molecules, *Nature*, 338: 137-139 (1989).
14. H. Böttcher, T. Fritz, and B. Vaupel, Fotochemische Anwendungen von Farbstoffaufdampfschichten, *Zeitschrift für Chemie*, 29:368-377 (1989).
15. H. Böttcher, Specific optical and electrical properties of vacuum-deposited thin films of dyes, *J. prakt. Chemie*, 334:14-24 (1992).

16. H. Böttcher, T. Fritz, and J.D. Wright, Fabrication of evaporated dye films and their application, *J. Mater. Chem.*, 3(12):1187-1197 (1993).
17. D. Wöhrle and D. Meissner, Organic solar cells, *Adv. Mater.*, 3: 129-134 (1991).
18. J.A. DeRose, T. Thundat, L.A. Nagahara, and S.M. Lindsay, Morphology of epitaxially grown gold on glass and mica, *Surface Science*, 256: 102-108 (1991).
19. C.C. Leznoff and A.B.P. Lever, eds., Phthalocyanines, Properties and Applications, VCH Publishers, Inc., New York (1989).

CHARACTERIZATION OF POLY(TETRAFLUOROETHYLENE) SURFACES BY ATOMIC FORCE MICROSCOPY— RESULTS AND ARTIFACTS

A. J. Howard,[1] R. R. Rye,[1] and K. Kjoller[2]

[1]Sandia National Laboratories
MS 0603 - PO Box 5800
Albuquerque, NM 87185-0603
[2]Digital Instruments, Inc.
520 E. Montecito St.
Santa Barbara, CA 93103

Abstract: The surfaces of virgin and chemically etched poly(tetrafluoroethylene) (PTFE) have been studied using atomic force microscopy (AFM) in both contact and tapping modes. While attempting to perform AFM in contact mode on this relatively soft polymeric material, tip-induced imaging artifacts (presumably due to blunt tips and tip-to-surface interactions) were identified when the results were compared to scanning electron microscopy (SEM) surface images. When subsequent AFM imaging was performed in tapping mode it was apparent that these tip-induced artifacts were eliminated. Comparable tapping mode AFM and SEM images were obtained for even the highly porous, unstable surface that results from sodium naphthalenide etching of PTFE. AFM imaging in tapping mode of virgin and etched PTFE surfaces shows the three-dimensional character of the etched surface necessary for mechanical interlocking and resultant strong adhesion.

INTRODUCTION

Due to its low dielectric constant (~2.0), PTFE is a substrate of interest in our laboratory for device applications that operate at very high frequencies or have capacitive coupling problems. Strong metal adhesion directly to PTFE is attainable by chemical etching the PTFE in a sodium napthalenide solution.[1] Chemical etching in this case refers to a specific reaction with the F atoms of PTFE, which produces a very porous, unstable, carbon-rich surface. Understanding how it is possible to get strong, reliable, and controllable adhesion to this well known nonstick surface is a major concern for scientists and engineers who are fabricating devices or PC boards and manufacturing products from etched PTFE and PTFE-like materials.

In previous surface science studies of etched PTFE we have determined that the surface roughness (measured area to geometric area ratio) is extremely high (200), based on BET adsorption isotherms, indicating that it is a porous, high energy surface.[2] Pull tests were also performed on Cu-coated, etched PTFE samples and the adhesive strength, 27 kg load failure, was found to be comparable to commercial Cu-coated PTFE (glass-fiber filled) (Duroid, 34 kg load failure, Rogers Corp.).[1] XPS analysis of the fracture surfaces from the pull test yield spectra that are nearly identical to virgin PTFE, which indicates that failure is not due to adhesive failure but near cohesive failure in the bulk PTFE.[3] Finally, RBS analysis of Cu coated etched PTFE shows Cu penetration and PTFE defluorination to a depth of ~3000 Å. Thus, the conclusion is that strong Cu adhesion is due to penetration deep into the etched PTFE surface, a process that results in strong mechanical interlocking.[4]

The common denominator to all of the surface studies of etched and virgin PTFE is that obtaining surface morphological information is central to understanding and controlling adhesion to etched PTFE. Consequently, we have investigated the surface morphology of etched and virgin PTFE using contact mode atomic force microscopy, tapping mode AFM, and scanning electron microscopy. The results and implications of this study will be discussed in this paper.

EXPERIMENTAL

All PTFE samples were cut into approximately 2.5 cm x 2.5 cm squares from the same 1.3 mm-thick commercially available sheet. The virgin white PTFE samples were degreased in an ultrasonic bath in acetone and methanol prior to etching. The samples were then etched at room temperature for 30 s using a sodium napthalenide solution (Poly-Etch, Matheson Gas Products or Tetra-Etch, W.L. Gore & Assoc.). Etching time for this ~3000 Å deep etch is not critical since this step is self-limiting, apparently due to the build-up of etch products.[5] The etched samples were then quickly rinsed in running water and then thoroughly rinsed sequentially for 5 min in ultrasonic baths of water, acetone, and methanol, and then air dried. Due to the highly reactive nature of the etched, now brown in color, PTFE surface, characterization steps were normally performed soon (< 24 hours) after etching. For SEM micrographs, the etched and virgin PTFE samples were coated with a ~100 Å conductive carbon coating.

The contact mode AFM images were taken using a Digital Instruments Nanoscope III, SA-AFMJ atomic force microscope operating in clean room-controlled air (40% R.H., 21° C). Standard Si_3N_4 pyramidal tips with ~500 Å end tip diameters were used for the contact mode imaging. For both modes of AFM operating, no additional sample preparation, other than etching and cleaning, was performed. The tapping mode AFM images were also taken using a Digital Instruments Nanoscope III, multimode atomic force microscope operating in normal room air. Si tips with ~100 Å end tip diameters and ~ 300 kHz resonant oscillating frequency were used for the tapping mode imaging. To get representative images for both contact and tapping mode, several 5 x 5 μm and 10 x 10 μm square areas were imaged at a resolution of 512 data points per line. A discussion contrasting these two types of AFM imaging will be given in the next section.

RESULTS AND DISCUSSION

As was stated previously, one of the main potential uses for PTFE is as a substrate for high frequency (~10 GHz) device applications; this is the driving force for PTFE-related research in our laboratory. We have previously developed three new fabrication processes for patterning microwave circuits directly on etched PTFE.[6] During the development of these processes, we experimentally observed intermittent problems with fine-line metal feature resolution and metal line edge roughness. These problems were later attributed to improper photolithographic patterning steps, but since surface roughness, which is necessary for strong adhesion, can be a potential problem in high resolution lithography, three-dimensional characterization of the etched surfaces is necessary.

We first investigated the surfaces of virgin and etched PTFE by SEM. Representative plan-view SEM images of these two surfaces are shown in Figure 1. The virgin samples have a moderately open surface structure with considerable filament-like features present (see Figure 1(a)) and several long, unidirectional scratches. The etched PTFE surface (see Figures 1(b), 1(c), however, has a high density of small cracks, all running ~perpendicular to the long scratches. At high magnification, see Figure 1(c), a fibrillar network is seen connecting the two sides of the cracks in the etched PTFE surface. For more details on these SEM characterization results of PTFE and other PTFE-like materials, like FEP, see Rye.[7] The SEM results are invaluable due to their great depth of field but unfortunately plan view SEM images are not calibrated for height.

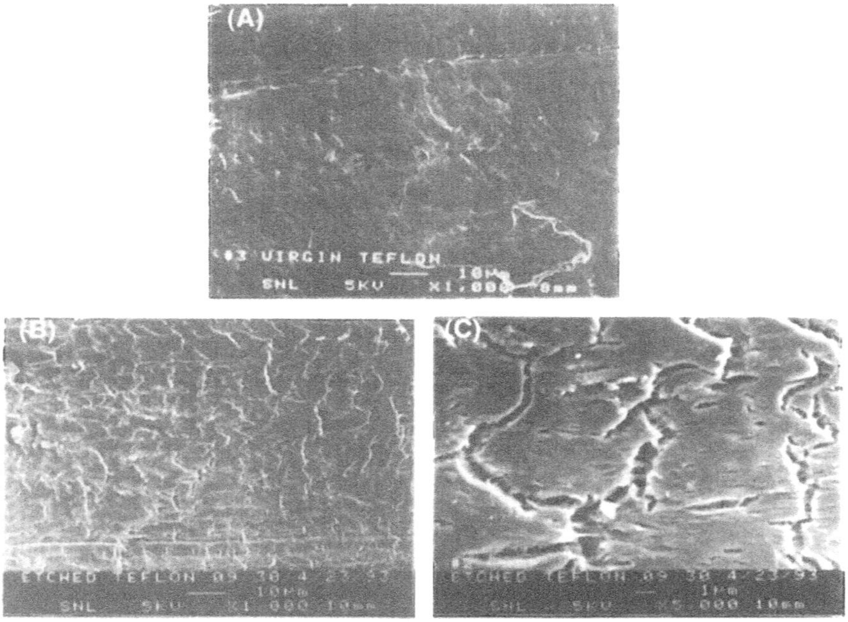

Figure 1. SEM images of (a) virgin and (b), (c) etched surfaces of PTFE.

To obtain more detailed morphological information, AFM imaging in contact mode (tapping mode AFM wasn't available yet at Sandia) was performed. Several attempts in contact mode yielded the same basic results; see Figure 2 for examples, in complete disagreement with the previous plan view SEM images. In all of the AFM images, changes in brightness indicate height differences such that the brightest regions have the maximum height and the darkest regions are low. It should be noted that during contact mode AFM imaging of PTFE it was extremely difficult to get stable, reproducible images; often the image appeared to shift from scan to scan.

The contrast between contact mode AFM and SEM of etched PTFE is particularly informative; SEM shows a high density of small cracks while AFM shows a surface dominated by square plates. These results suggest that since the PTFE surface is extremely soft, and the

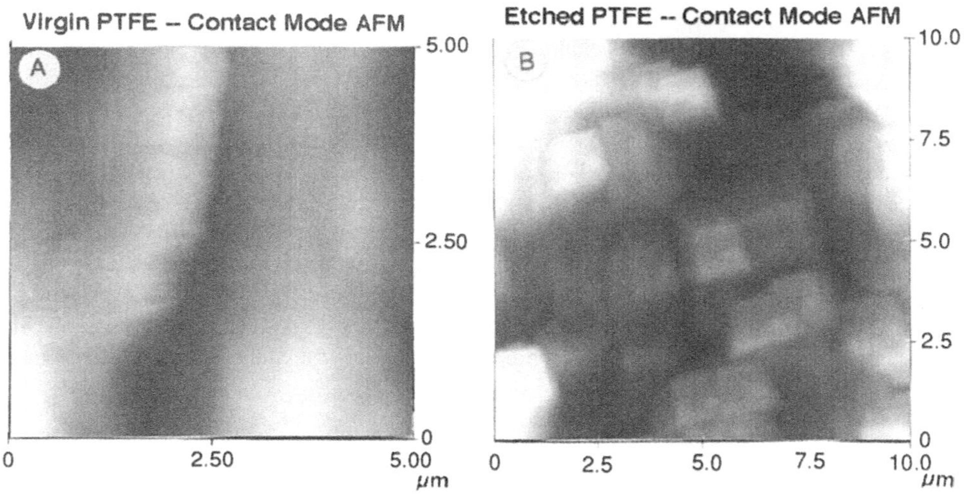

Figure 2. Contact mode AFM images (top view) of (a)virgin and (b) etched PTFE.

etched PTFE surface even worse, tip-to-surface interactions were causing our images to be dominated by artifacts. An SEM image of our blunt tip is in Figure 3. The only clear correlation that can be made from these images is between the square plate structure observed for etched PTFE in contact mode AFM and the square, blunt nature of the pyramidal tip. (Apparently, etched PTFE is so soft attempts at imaging reflect the shape of the AFM tip, not the surface.)

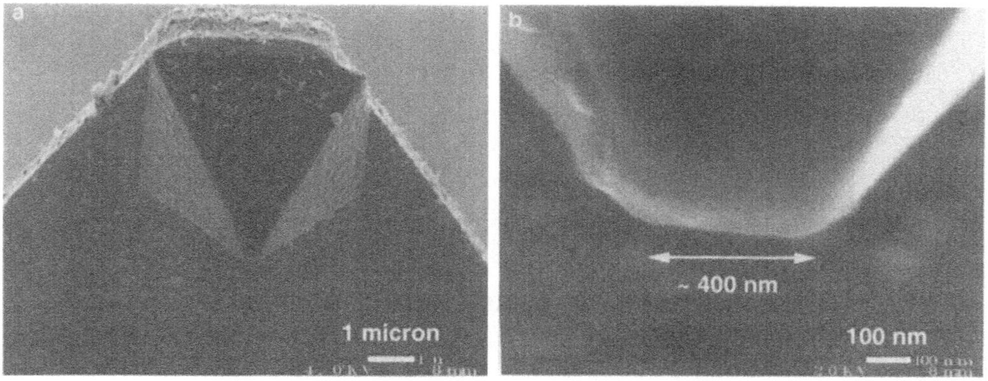

Figure 3. SEM micrographs of a "blunt" contact mode AFM tip. (a) Low magnification view of the tip; (b) higher magnification view of the tip.

Our results are in agreement with the observations made by Grutter et al.[8] when imaging even hard diamond thin films; they noted that AFM imaging with blunt tips can cause severe misrepresentation of the true surface morphology. Subsequent attempts at contact mode AFM of PTFE with oxidatively sharpened Si tips, end tip radius ~100 Å, also yielded images which contained artifacts (streaks) and also did not agree with our SEM results. Our conclusions at this point were: (1) due to the soft nature of PTFE, imaging it is a severe test for an AFM, and (2) since contact mode AFM operates with continuous surface contact at typical force loads of 100 nN, it is highly unlikely that any contact mode AFM images of PTFE would be meaningful. Consequently, after discussing our dilemma with Digital Instruments, they agreed to attempt tapping mode AFM on our virgin and etched PTFE samples.

In tapping mode (see Zhong et al.[9] for more details) the cantilever/tip oscillates at its resonant frequency (~300 kHz) with an oscillation amplitude of 20 to 100 nm. Hence, the tip has only brief intermittent contact with the surface at a force load estimated to be roughly 0.2 to 5 nN.[9] This combination of intermittent contact at lower force loads and large oscillation amplitudes is, as we have discovered, ideal for imaging soft materials such as PTFE in a non-destructive, artifact free fashion. Typical tapping mode (plan view) AFM images of virgin and etched PTFE, which were taken at Digital Instruments in a blind test fashion, are shown in Figure 4. Note the striking resemblance of the tapping mode images, even down to the detail of the fibrillar network, shown in Figure 4 to the SEM images shown in Figure 1. It is clear that with tapping mode, it is possible to obtain meaningful 3-D surface morphological information on PTFE.

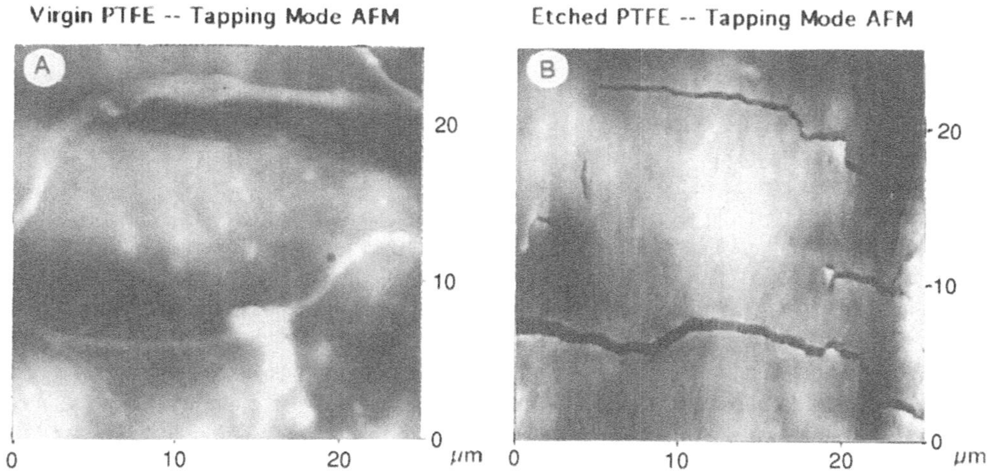

Figure 4. Tapping mode AFM images (top view) of (a) virgin and (b) etched PTFE. Note the fine fiber structure across the cracks in the etched surface.

Now that we have confidence that the tapping mode AFM is giving us believable morphological information, 3-D views of the surfaces of virgin and etched PTFE are analyzed. Representative 3-D surface plots are shown in Figure 5. It is interesting to note when comparing the two surfaces that the "hills" in the virgin PTFE surface, which are undoubtedly associated with skiving-induced stress, appear to open up and result in the fiber-connected cracks, which are prevalent in the images of the etched surface. Further line-by-line cross-sectional analysis of these images shows that the heights of the "hills" and the depths of the cracks are comparable (~2000 to 8000 Å). So, finally by combining this AFM 3-D information with previous results, a complete explanation seems to be falling into place. Stress causes the "hills" in virgin PTFE, which subsequently turn into cracks after chemical etching, and these cracks make it possible to get strong adhesion to etched PTFE through mechanical interlocking.

These tapping mode AFM results are in agreement with a previously published model that explains how it is possible to get strong metal adhesion to etched PTFE (see Rye[7] for more details). Basically, PTFE can not be melted without decomposition and no solvent exists that is capable of dissolving PTFE. As a result, thick sheets of commercial PTFE are produced by

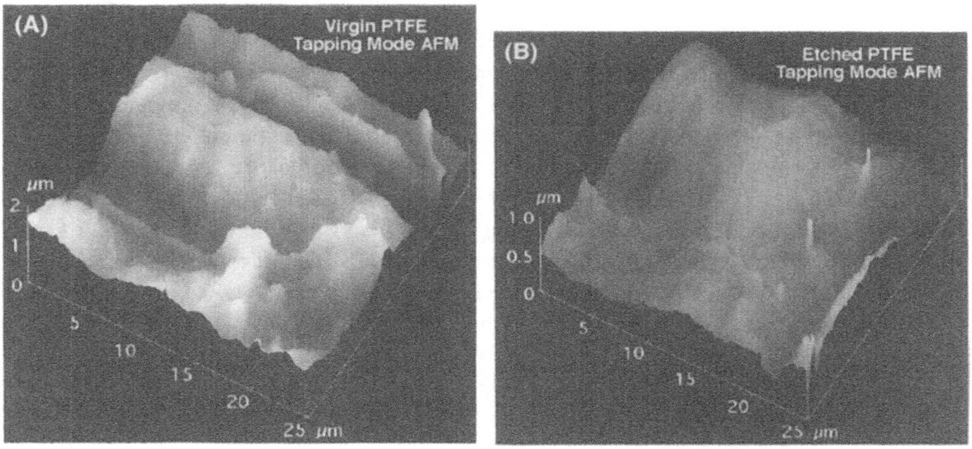

Figure 5. Three-dimensional tapping mode AFM images of (a) virgin and (b) etched PTFE.

hot pressing from the powder, then direct pressing into thick sheets. Thin sheets, required for proper microwave device performance, that are less than 0.64 cm are skived from hot pressed cylinders.[10] Skiving is a process (similar to the one used for making plywood from logs) by which thin sheets are sliced from a rotating, hot-pressed, PTFE cylinder. Knowing that thin PTFE is produced from the skiving process explains why the long macroscopic scratches exist; inevitable nicks and imperfections in the knife edge are transferred to the skived PTFE. Also, the skiving process is expected to produce stress in the thin sheets which would be concentrated in an orientation parallel to the knife edge (or perpendicular to any macroscopic scratches). The chemical etching process apparently relieves this stress leading to fiber-connected cracks. These cracks allow for deep penetration of the electrolessly deposited Cu thin film, which results in strong mechanical metal-to-PTFE interlocking and consequently strong metal adhesion.[7]

This model is in agreement with previous failed attempts at metallizing other smoother forms of PTFE and FEP, which are produced directly from the melt and do not have the built-in stress that is necessary to form etchant-induced mechanical interlocking.[7] This information has had a major impact on our microwave device fabrication on PTFE program as it clearly points out: (1) our limitations in substrate selection; (2) the need to perform in-line, nondestructive substrate roughness inspection prior to fabrication; and (3) that resolution and reliability improvements will depend on generation of a virgin PTFE substrate with a higher degree of stress uniformity.

CONCLUSIONS

We studied virgin and sodium napthalenide etched surfaces of PTFE using atomic force microscopy (AFM) in both contact and tapping modes. Contact mode AFM imaging on this relatively soft polymeric material is essentially impossible as images were in gross disagreement with SEM images. Tip-induced imaging artifacts, initially due to blunt Si_3N_4 tips and subsequently due to sharp Si tip-to-surface lateral force interactions showing up as streaks due to a lack of frictional force control, were identified as the problem. When AFM imaging was performed in tapping mode these tip-induced artifacts and deformation due to lateral and normal forces were eliminated as images were in perfect agreement with SEM results. Metal patterning with strong adhesion directly to PTFE for high frequency microwave device applications is attainable by etching the PTFE in a sodium napthalenide solution. AFM imaging in tapping mode of virgin and etched PTFE surfaces has provided three-dimensional surface information, which is in agreement with a previously published model.[7] This model states that strong metal adhesion to this nonstick surface stems from mechanical interlocking; deposited metal penetrates into deep stress-relieved cracks, which are associated with the thin (<0.64 cm) PTFE sheet fabrication process.

This information, which complements previous surface science investigations of PTFE-like materials, has had a major impact on our microwave device on PTFE program. It has pointed out PTFE substrate selection rules, demonstrated the need to perform prefabrication substrate roughness acceptance inspections, and indicated a path for potential resolution and reliability improvements. Experiments are in progress to demonstrate improvement and also to further understand exactly why tapping mode is successful in imaging this soft, high energy, polymeric material.

ACKNOWLEDGMENTS

This work performed primarily at Sandia National Laboratories is supported by the U.S. Dept. of Energy under contract DE-AC04-94AL85000.

REFERENCES

1. R. R. Rye and A. J. Ricco, Patterned adhesion of electrolessly deposited copper on poly(tetrafluoroethylene) *J. Electrochem. Soc.*, 140, 1763-1768 (1993).
2. R.R. Rye, Radiation hardening of poly(tetrafluoroethylene) against chemical etching, *J. Polym. Sci.: Polym. Phy.*, 26, 2133-2144 (1988).
3. R.R. Rye, Spectroscopic evidence for radiation-induced crosslinking of poly(tetrafluoroethylene) *J. Polym. Sci.: Polym. Phys.*, 31, 357 (1993).
4. R. R. Rye, G. W. Arnold and A. J. Ricco, Characterization of the copper poly(tetrafluoroethylene) interface, *J. Electrochem. Soc.*, 140, 3233-3239 (1993).
5. G.W. Arnold and R.R. Rye, Ion beam analysis of the effects of radiation on the chemical etching of poly(tetrafluoroethylene) Nuclear Instruments and Methods B (Proceedings of Radiation Effects in Insulators - 5), B46, 330-333 (1990).
6. A.J. Howard, R.R. Rye, A.J. Ricco, D.J. Rieger, M.L. Lovejoy, L.R. Sloan, and M.A. Mitchell, New methods for circuit fabrication on poly(tetrafluoroethylene) substrates, *J.Electrochem. Soc.*, 141, 3556 (1994).
7. R. R. Rye, Surface stress-dependent adhesion to fluorinated polymers, *J. Polymer Sci.: Polym. Phy.*, 32, 1777-1785 (1994).
8. P. Grutter, W. Zimmermann-Edling, and D. Brodbeck, Tip artifacts of microfabricaated force sensors for atomic force microscopy, *Appl. Phys. Lett.*, 60, 2741-2743 (1992).
9. Q. Zhong, D. Inniss, K. Kjoller, and V.B. Elings, Fractured polymer /silica fiber surface studied by tapping mode atomic force microscopy, *Surf. Sci. Lett.*, 290, L688-L692 (1993).
10. Private communication, C. Crowley, Regal Plastics, Houston, TX.

SCANNING PROBE MICROSCOPY STUDIES OF ISOCYANIDE FUNCTIONALIZED POLYANILINE THIN FILMS

Timothy L. Porter[1]* and Andrew G. Sykes[2]
[1]Department of Physics
Northern Arizona University
Flagstaff, Arizona 86011
[2]Department of Chemistry
Northern Arizona University
Flagstaff, Arizona 86011
**Correspondence addressee*

Abstract: We have used the technique of scanning probe microscopy to study the nanometer-scale structure of thin films of polyaniline and isocyanide functionalized polyaniline. Structural changes on the film surfaces were studied as a function of protonation in aqueous HCl, as well as exposure to Ir^+ cations in solution. Film growth of pure polyaniline is in a layer-by-layer mode, with an intricate, nanometer-scale domain structure. Ir^+ incorporation leaves this structure intact. Electropolymerized isocyanide functionalized films exhibited a highly oriented fibrous structure, with individual strands averaging 25 Å in diameter. Protonation of these films resulted in aggregation of these strands into elongated bundles of average diameter 200 Å. Incorporation of Ir^+ cations into the film matrix produced an oriented, interlocking structure of polymer bundles of diameter 400 Å. We also discuss the relationship of the microscopic structure of these films to their macroscopic conductive properties.

INTRODUCTION

Since the discovery that the normally insulating polyacetylene could be "doped" into a conductive state, many other synthetic metal polymeric materials have been prepared and studied. Certain of these conducting polymer films may be functionalized by incorporation of various chemical species during polymerization. This results in the ability to synthesize materials with a wide range of desired chemical, electrical, or physical behaviors. Of these polymeric materials, polyaniline has proved to be highly adaptive in terms of the number of physical, chemical and electrical states accessible in the laboratory. Polyaniline may be prepared in a number of different nonconducting oxidation states, one of which (the "emeraldine" oxidation state) may be transformed into a conductive state by adding or subtracting charge from the polymer chain. Preparation in or post polymerization exposure to

Atomic Force Microscopy/Scanning Tunneling Microscopy 2
Edited by S.H. Cohen and M.L. Lightbody, Plenum Press, New York, 1997

acids will result in positively charged mobile carriers being added to the polymer chains. This "protonation" of the polymer results in a material with macroscopic conductivities in the range 1-5 S/cm. Further post polymerization processing, such as heat assisted film stretching may result in even greater conductivities.[1] Also, substituents may be incorporated into polyaniline that alter any number of the chemical and electronic properties.[2] Polymerization of aniline in the emeraldine oxidation state generally results in two initial material types, depending on the precise method of formation.[3] Films or bulk powders prepared in the conducting, acid-doped form have been reported to be semicrystalline in structure, with crystalline domain sizes reported in the range of 30 - 200 Å.[3] Polyaniline prepared initially in base form may be either semicrystalline in powder form or amorphous when prepared as films or fibers.

We have also studied the incorporation of Ir^+ in polyaniline and 3-isocyanoaniline thin films. It has been shown that isocyanide complexes of Rh(I) and Ir(I) catalyze photochemical hydrogen atom transfer reactions, and also abstract halide atoms from environmentally toxic halocarbons. Isocyanide d^8 metal complexes also form adducts with both Lewis acids and bases, and catalyze photoevolution of dihydrogen in aqueous acid solutions. Synthesizing conductive polymer films incorporating isocyanide functionalities could act to bind metal ions, and manipulation of the local metal environment through the conductive matrix could be important in the control of the catalytic reactions mentioned above. Also, new electrode materials with applications as ion sensors and metal recovery or release systems may be possible.

EXPERIMENTAL

Polyaniline films were prepared in their conductive state electrochemically from a solution of 0.1 M aniline in 0.1-0.5 M $HClO_4$ using standard cyclic voltammetric techniques. The potentials were cycled between -0.5 to +1.2 V. All films were plated on Pt foil surfaces, with film thicknesses greater than 1 μm in order to obscure any substrate structural features. These films were radiant blue in color, and generally free of defects. Ir^+ incorporation was accomplished by exposure to 1-2 mM $[Ir(COD)Cl]_2$ (COD=cyclooctadiene) in chloroform for 12-24 hours. For the isocyanide functionalized films, 3-isocyanoaniline powder was synthesized by first dissolving 30 grams of 1,3-phenylenediamine in 400 mL chloroform. To this, 200 grams of potassium hydroxide flake dissolved in 800 mL H_2O/ 120 mL ethanolmixture was added. After two hours under reflux, the reaction mixture was cooled to room temperature and the water layer was separated from the chloroform layer and extracted using additional chloroform. The combined chloroform layers were washed with distilled water and then dried over anhydrous sodium sulfate powder. Chloroform was evaporated under reduced pressure and a dark, tarry residue was obtained. This residue was then extracted with ligroine, and the combined crystalline material was again recrystallized with hexane. This resulted in a pale, yellow solid material. Infrared, ^1H NMR and mass spectroscopy successfully characterized the product. Anodic polymerization of this material was then performed in 0.1 M $NaClO_4$/MeCN solution. Films were plated on Pt foil surfaces, with film thicknesses again greater than 1 μm in order to obscure any substrate structural features. These films were yellow-gold in color, free of defects such as pinholes, and non-electrochromic with variation in potential (as polyaniline itself).

All scanning probe images of the film surfaces were obtained with an Auto-Probe CP from Park Scientific Instruments. The cantilevers used for contact imaging were of force constant 0.03 N/m, resonant frequency 19 kHz, and 180 μm in length. The actual scanning force used for contact imaging was less than 100 nN in all cases in an attempt to minimize the distortion of the polymer surface features for this mode. For both contact and noncontact images, highly sharpened conical tips were used. In noncontact mode, the cantilevers were of force constant 12 N/m, with theoretical resonance frequencies of 260 kHz. Scanning speeds ranged from 1 μm/s for larger areas to less than 200 Å/s for higher resolution scans. All non-contact scans were repeatable, indicating little or no damage to the polymer surfaces during scanning.

RESULTS AND DISCUSSION

Figure 1 shows an image taken of the Ir^+ exposed polyaniline surface in the normal contact mode of the SPM. The image in Figure 1 shows a surface region of 100 μm^2. In this contact imaging mode, the SPM cantilever is brought sufficiently close to the surface so that a repulsive force is exerted between the cantilever tip and the surface. This "contact" between the tip and surface is usually in the range 10^{-9} - 10^{6} N, and results in the bending of the cantilever (and possible distortion of the surface). In a typical constant force scan, the vertical position of the cantilever mount is continuously modulated while the tip is rastered over the surface in order to maintain some preset constant repulsive force between the tip and surface. The image in Figure 1 was obtained at a constant force of only 50 nN. Even at this low force, the images appear featureless and blurry. In the figure, numerous horizontal streaks indicate

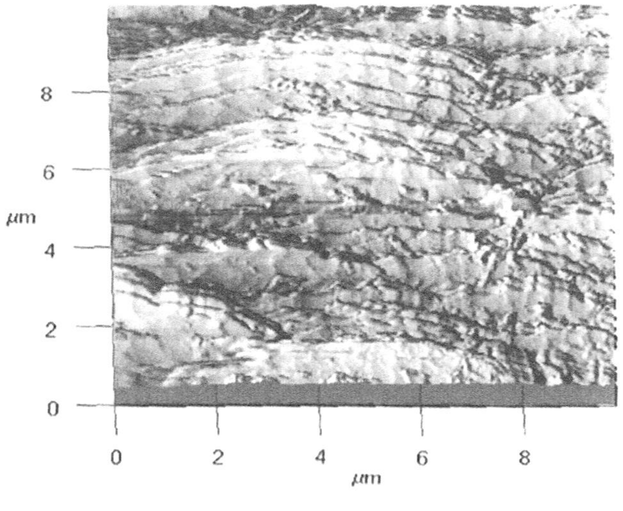

Figure 1. 100 μm^2 contact mode SPM scan on surface of Ir^+ exposed polyaniline.

that the SPM cantilever may have been physically dragging surface material as it was scanned back and forth. In some cases with functionalized conducting polymers (such as 3-isocyanoaniline later in this report) the material is sufficiently stiff to allow for high resolution contact mode SPM imaging. These unsubstituted polyaniline films, however do not lend themselves to study with this technique. In Figure 2 we show an SPM image of the same film surface taken in the SPM noncontact mode. The dimensions of this image are also 100 μm^2, and the vertical range is 5000 Å from light to dark. Comparison of this image with that in Figure 1 shows that in this scanning mode, the surface material is not distorted during the scan. In this noncontact imaging mode, the cantilever tip is rastered across the sample surface at a separation such that the interaction force between the tip and surface is the small, attractive van der Waals force. The tip-sample separation in this mode is user controllable, and usually in the range of 5 - 100 Å. This weak attractive force is generally in the pico-Newton range, which results in little or no surface deformation during scanning, even for very soft surfaces. In order to measure the tiny cantilever deflections in this mode, the cantilever itself is vibrated at one of its resonant frequencies (200-300 kHz). As the tip is moved over the surface, changes in the surface topography result in a constantly changing tip-surface distance.

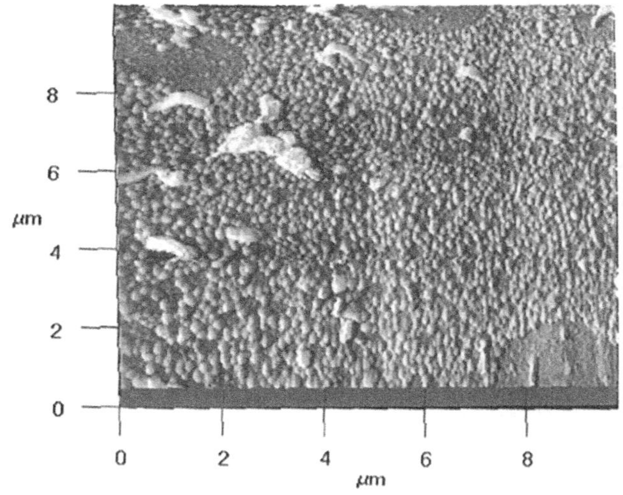

Figure 2. 100 μm^2 SPM noncontact image of same surface in Figure 1.

This fact, in turn, affects the force field gradient that the cantilever tip is vibrating in, slightly altering the tip vibration frequency. It is this frequency change that is measured by the instrument, and the feedback circuitry then readjusts the tip-sample separation to bring the frequency (and thus the separation between tip and sample) back to the desired point. While this method of measuring surface topography has very distinct advantages for imaging soft surfaces, the larger tip-sample separations used as well as the generally flatter (shallow gradient) force field in this attractive region may result in lower attainable lateral resolutions. The use of very sharp, conically shaped SPM tips as opposed to more common pyramidal tips helps to overcome these problems, however.

In Figure 3, an SPM noncontact image of dimensions 4 μm^2 is shown for the same surface. The vertical range in this image is 4000 Å. In this image, bundles from both the topmost and second layer are imaged. The average size of these bundles is 1000 ± 75 Å. The layer-by-layer growth of these polymer bundles is very likely due to the electrochemical plating process used to grow these films, and is not observed in chemically prepared, free-standing polyaniline thin films.[4] Polyaniline films that were not exposed to $^{+}$Ir displayed the same nanometer-scale bundle structure as observed in these images. The average polymer bundle size in the unexposed films was also the same. While Ir^{+} incorporation into the exposed films has been confirmed,[5] little or no structural effects on this scale have been observed.

Theories for the macroscopic transport of charge in polyaniline have generally fallen into two categories: the granular conductor models, and the Fermi glass conductor models.[6] In the granular conductor models, macroscopic conductivity is limited by charge carrier "hopping" between highly conducting islands or bundles which may be embedded in a poorly conducting matrix. The size of these bundles or islands has been estimated to be about 200 Å. Each bundle or island contains polymer strands with sufficient crystalline order so that electron

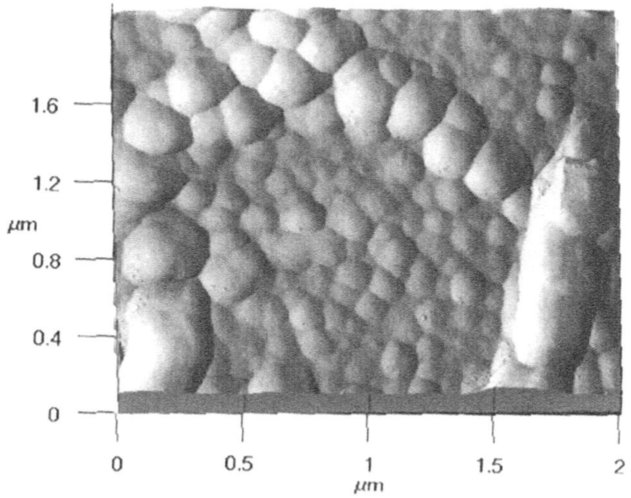

Figure 3. Noncontact image of 2 μm^2 region on same surface as Figure 1. Bundles from both the topmost and second layer are visible. The vertical range is 4000 Å from bright to dark. The average dimension of the polymer bundles is 1000 ± 75 Å.

wavefunctions are three-dimensionally delocalized, resulting in metallic behavior within each bundle. In the Fermi glass models, the material structure is composed of an entirely amorphous or disordered array of polymer strands, each of which conducts along its length. Macroscopic film conductivity is then limited by the charge carrier hopping that must occur between individual polymer strands for the material to conduct on the macroscopic scale. It is seen from these images that indeed these polyaniline films are composed of small polymer bundles, and not just an amorphous array of polymer chains. This observation is possible for us only because of the noncontact SPM operational imaging mode. The observed size of these bundles is 1000 Å, greater than that proposed for the granular conductor model, however. One explanation for this may be that individual bundles are not entirely well ordered, but consist of crystalline or semicrystalline cores that themselves behave like metallic conductors. These cores are then encapsulated in amorphous, but partially conductive sheaths, that serve to limit the overall film conductivity. This model would also explain the difficulty in obtaining direct images of the polymer bundle crystalline structure with techniques such as scanning tunneling microscopy (STM).

In a previous study,[5] it was clearly demonstrated that electropolymerization of aniline containing the strong electron-withdrawing isocyanide group 3-Isocyanoaniline in nonaqueous

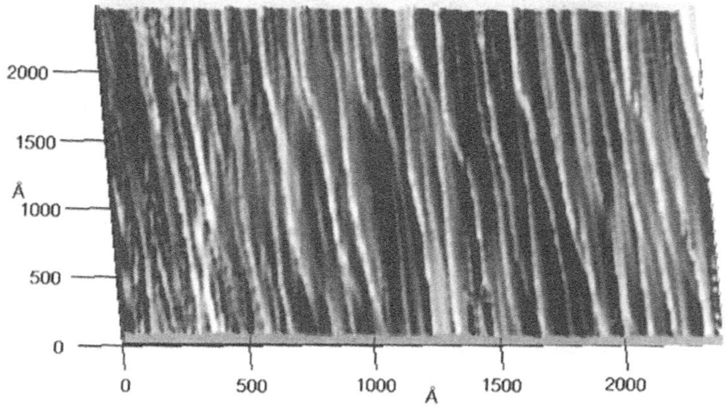

Figure. 4. High resolution SPM image of same film surface as in Figure 1. Individual polymer fibers or strands arranged in a close-packed structure are clearly visible. The average diameter of these strands is 25 ±5 Å.

media was possible. It was also demonstrated that these films readily coordinate Ir cations from solution.

In Figure 4 a high resolution scan on the 3-isocyanoaniline surface of dimensions 2500 $Å^2$ is shown. The strand-like microstructure of the film is clearly evident. The measured average diameter of the strands is 25 ± 5 Å. This strand diameter is in good agreement with other measured functionalized aniline strand dimensions.[7] In this previous study, individual strands of electropolymerized poly-hydroxyaniline were imaged using the technique of scanning tunneling microscopy (STM). The diameter of these strands or coils was measured to be 20 ± 5 Å. It was proposed that these strands are simple helical coils, with periodicity along the coil of about 30 Å. In this study, the diameter of individual coils is readily measurable, but any periodicity along the length of the strands is not seen. This may indicate that these are simply thin aggregates of linear polymer chains rather than helical coils of single polymer chains.

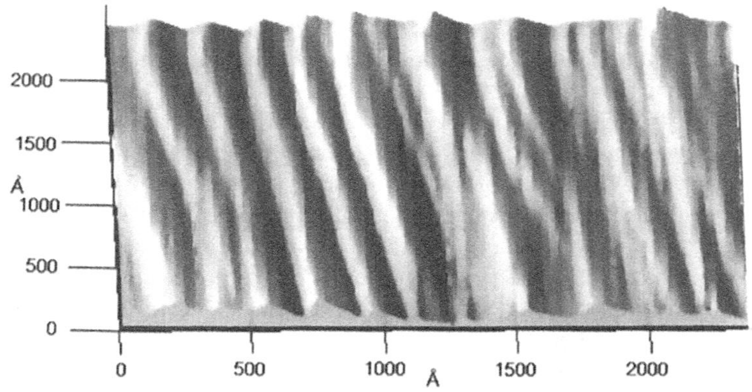

Figure 5. Surface of isocyanide-functionalized aniline after protonation in aqueous HCl. The fibrous structure of the film has given way to more of an island or bundle structure. These bundles are still partially oriented in the original direction of film orientation. The average diameter of the bundles is 200 Å.

Also of interest is the apparent degree of ordering in these films. In a study using the techniques of near x-ray absorption fine structure (NEXAFS) and infrared reflection-absorption spectroscopy (IRRAS), it was shown that initial layers of 3-methylthiophene would grow with preferential orientation parallel to the substrate (Pt). This was due to the π-interaction with the metallic substrate. Ordering would then be imposed on subsequent film layers, with the degree of orientation decreasing as film thickness increases. In the present study, the higher than expected film orientation may be due to the isocyanide-Pt interactions in solution. Isocyanides display strong σ donation and extensive back acceptance of π-electrons from metals in low oxidation states. This interaction between the isocyanide group and the Pt substrate may not only enable efficient electrooxidation of the monomer but may also result in orientation of the first few polymer layers. Subsequent layers may grow in a template fashion, presumably with decreasing orientation and order as film thickness increases.

Isocyanide-functionalized aniline films were also protonated by exposure to 1 M HCl in solution. Total exposure time was 24 hr. The films were then dried for 24 hrs. prior to SPM imaging. Figure 5 shows a 2500 $Å^2$ SPM image of the film surface after protonation. The region shown here is characteristic of the entire film surface structure, which is not segregated into individual grains. The fibrous nanostructure of the unprotonated film has been completely replaced by an island or bundle structure. This type of reorganization upon protonation is typical of polyaniline and polyaniline-based compounds, and is a result of incorporation of protons into the polymer chains as well as the added presence of Cl counterions in the polymer matrix.

For pure polyaniline films or fibers, this EB-II to ES-II transition is also accompanied by a reordering of the polymer microstructure into a crystalline or semicrystalline state. In this model, each polymer bundle possesses sufficient crystallinity to allow for three-dimensional electron wavefunction delocalization, resulting in a polaronic lattice electronic structure. The overall macroscopic conductivity of the material is, however, still limited by poor charge transport between individual polymer bundles. The dimension of the individual bundles observed on the isocyanide-functionalized films is on the order of 200 Å laterally and 2500 Å along the length of individual bundles. While the lateral bundle size is in good agreement with earlier studies on aniline and hydroxyaniline films, the bundles in the present study are considerably more elongated. We feel that this result is due primarily to the increased structural order present in the isocyanoaniline films prior to protonation. In contrast, pure polyaniline films exhibit little or no structural coherence on the nanometer scale prior to protonation, but order themselves into cigar-shaped bundles of average dimension 250 Å after protonation.

Unprotonated isocyanide-functionalized aniline films were also exposed to Ir^+ in solution. A 1-2 mM solution of $[Ir(COD)Cl]_2$ (COD = 1,4-cyclooctadiene) in CH_2Cl_2 was degassed with dinitrogen and an electropolymerized isocyanoaniline film was then immersed. This exposure of the functionalized polymer to the Ir^+ complex in solution binds metal ions to the modified surface, as with polyaniline. Metal ions typically form complexes with isocyanides through coordination of the isocyanide carbon which acts as a lewis base. Figure 6 shows a high resolution scan of a region 2000 $Å^2$ in dimension of this film. It is evident from this image that incorporation of the Ir^+ cations into the polymer matrix has resulted in structural reconstruction of the film surface morphology. Replacing the original fibrous structure is a structure consisting of interlocking, needle-like polymer bundles. The average length of the individual bundles is 1.25 ± 0.25 μm, in accordance with the original length of the polymer fibers. The average width of these bundles is now 400 ± 50 Å, however. Groups of individual fibers have coalesced laterally into these larger structures. The internal structure of these new needle-like bundles is sufficiently close-packed and uniform that SPM imaging on a smaller scale could not be achieved. These isocyanide functionalized films exhibit important properties in regards to applications as ion sensors, metal recovery systems, and control of certain catalytic reactions involving environmentally toxic halocarbons. A study of the microscopic structure of these materials and the relationship between this structure and the electronic and chemical properties of the films is needed in order to understand how these materials may be used to greater advantage in these applications. Further studies of these structural effects on aniline as well as functionalized aniline films are currently underway.

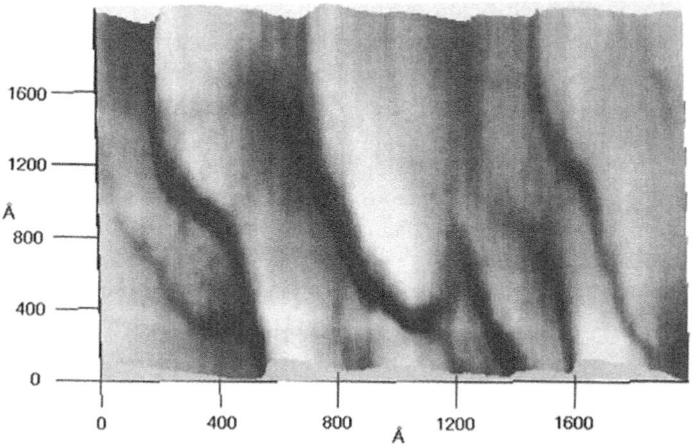

Figure. 6. High resolution scan of the Ir exposed film. The fibrous structure has given way to a film structure consisting of interlocking polymer bundles. These bundles are highly elongated, and partially oriented in the original film orientation direction. The average width of these bundles is 400 ± 50 Å

CONCLUSION

Noncontact SPM data on both pure and Ir^+ exposed electrochemically prepared polyaniline films show that the material is segregated into small, nanometer-scale bundles. The average dimension of these approximately spherical bundles is 1000 ± 75 Å. SPM data support a granular conductor model for this material, in which macroscopic charge transport is limited by poor conduction between otherwise highly conductive bundles. Films of isocyanide-functionalized polyaniline were also polymerized using standard electrochemical methods. SPM scans on unprotonated, as-polymerized isocyanoaniline films reveal a nanometer-scale fibrous structure. The fiber diameters average 25 ± 5 Å, and their length 1.25 ± 0.25 μm. The fibers show a large degree of orientation, due to large isocyanide-Pt interactions in the electrochemical solution. After protonation, the film material has coalesced into a nanometer-scale bundle structure similar to that seen in aniline and other functionalized aniline films after protonation. Exposure to Ir^+ cations in solution also results in a structural reconstruction for the films. A large scale system of needle-like polymer bundles is observed. These bundles are highly interlocked, and have lengths similar to the fibers in the original films. The width of these new structures is measured to be 400 ± 50 Å.

ACKNOWLEDGMENTS

This work was supported by the NSF (DMR-9217525), Research Corporation, the Vice President for Research at NAU, and the Camille and Henry Dreyfus Foundation.

REFERENCES

1. P. Adams and A.P. Monkman, Observed anisotropies in stretch oriented polyaniline, *Synthetic Metals*, 41:627-633 (1991).
2. G. Caple, C.Y Lee, T.L. Porter, and P.I. Oden, Surface structural study of poly-hydroxy- aniline and aniline-(3-aminophenylboronic acid) co-polymer films, *J. Vac. Sci. Technol.* A10(4):606-610 (1992).
3. A.J. Epstein, M.E. Jozefowicz, A.G. MacDiarmid, J.P. Pouget, X Tang, X-ray structure of polyaniline, *Macromolecules*, 24:779-789 (1991).
4. K. Caple, G. Caple, T.L. Porter, Surface structure of free-standing polyaniline thin films, *Synthetic Metals*, 60:211-214 (1993).
5. G. Caple, T.R. Dillingham, T.L. Porter, Y Shi, A.G. Sykes, Electropolymerized films of isocyanide functionalized anilines: coordination of a heavy metal cation, *J. Electroanal. Chem.*, 380:139-145 (1995).
6. A.J. Epstein, C. Li, A.G. MacDiarmid, E. M. Scherr, Z.H. Wang, Three dimensionality of metallic states in conducting polymers: polyaniline, *Physical Review Letters*, 66:1745-1748 (1991).
7. G. Caple, C.Y. Lee, T.L. Porter, B.L. Wheeler, Scanning tunneling microscopy studies of substituted polyaniline thin films, *J. Vac. Sci. Technol.*, A9(3):1452-1456 (1991).

NEW DEVELOPMENTS IN AFM/STM

Moderator: Jose Perez, University of North Texas

INVESTIGATIONS ON THE TOPOGRAPHIC AND SPECTROSCOPIC IMAGING BY THE SCANNING TUNNELING MICROSCOPE

M. Hietschold,[1] O. Pester,[1] D. Porezag,[1] M. Röder,[1] H. Sbosny,[1] K.Walzer,[1] and L.Koenders[2]

[1]Solid Surfaces Analysis Group
Institute of Physics
Technical University of Chemnitz-Zwickau
Germany
[2]Physikalisch-Technische Bundesanstalt Braunschweig
Germany

Abstract: We present theoretical and experimental results on the imaging capabilities of the scanning tunneling microscope (STM). First, we report on the metrological application of the STM to defined probing of surface profiles on a nanometer scale. For a well-defined tip shape, a systematical quantitative analysis can be made to obtain quite accurate measurements, at least in some cases. The second example we consider is organic molecules imaged with submolecular resolution at ambient conditions. Pictures obtained experimentally are interpreted by Hartree-Fock calculations for (oxy-)cyano-biphenyl-alkanes and for metal-phthalocyanines. Moreover, new spectroscopic results - only accessible under ultrahigh vacuum (UHV) - are predicted theoretically.

INTRODUCTION

Presently, we are witnessing a development of the scanning tunneling microscope from a simple, qualitative imaging device towards a more quantitative measuring apparatus. In this way, the STM might become an important tool for metrology on a nanometer scale.

On the other hand, such a development strongly demands a better understanding of the imaging process itself on the corresponding scale. This fact is of special interest since there is an interplay of classical and quantum mechanical effects in the nanometer range. This interplay means especially that

- one has to answer exactly the question of what we are seeing,
- one has to study "artifacts" occurring in the images due to the imaging process (and how to prevent them), and
- one has to investigate what can be seen at all.

The central question is whether there is a unique relation between the image and what is called "reality." Such research should finally help to answer questions like: How can an image be optimized? What can we learn from a special image?

Theoretical modeling and comparative experiments will be the two only possible methods to deal with these problems. In this paper, we study corresponding problems for nanometer-scaled geometrical surface profiles and for the case of subnanometer resolution and spectroscopic imaging of organic molecules. Some first results obtained are presented in the proceedings of the previous symposium.[1]

TRANSFER-HAMILTONIAN FORMALISM

Within the standard model of Tersoff and Hamann[2] the tunnel current I_T between two electrodes 1 and 2 is given by

$$I_T \sim e \sum_{i,k} M_{ik} f_1(\epsilon_i) [1 - f_2(\epsilon_k - eU_T)]$$

where M_{ik} is the transition matrix element for tunneling between the states I and k, and f is the Fermi distribution function for both of the electrodes 1 and 2, respectively. U_T is the tunnel bias applied externally. The transition probability is determined under the assumption of an adiabatic perturbation that means a slowly "switching on" of the second electrode[3]

$$M_{ik} = 2\pi / \hbar \ |< H'_{ik} >|^2 \ \delta (\epsilon_i - \epsilon_k + e U_T)$$

with

$$H'_{ik} = - \hbar^2 / 2m \ \sum_j \int d\vec{S} \ [\psi_i^{(1)*} \nabla_j \psi_k^{(2)} - \psi_k^{(2)} \nabla_j \psi_i^{(1)*}].$$

This expression depends only on the unperturbed electron states of the single electrodes. The integration is over a surface between the electrodes.

Within the standard model[2] (spherically shaped tip electrode with only s-states, and a plane sample surface exhibiting the lattice periodicity of the corresponding crystallographic plane) the electron state wave functions of both of the unperturbed electrodes are known explicitly and the tunnel current can be calculated analytically leading to

$$I_T \sim \sum_{k_\|} | \psi_{k_\|}^{(2)} (\vec{r}_0) |^2 \ \delta (\epsilon_{k_\|} - e U_T - \epsilon_F^{(2)})$$

$$= \rho^{(2)} (\vec{r}_0, \epsilon_F^{(2)} + e U_T)$$

which means that the STM tip follows in the constant current mode of STM operation a contour of constant local density of states (LDOS) of the sample ρ_{sample} at the (fictitious) position $\vec{r}_0$ of the radius of curvature of the tip electrode:

$$I_T = \text{const.} \ \Rightarrow \ \rho_{sample} (\vec{r}_0, \epsilon_F^{(2)} + e U_T).$$

This model does not include:

- constraints due to geometrical restrictions in sample-tip shape configuration which turn out to be essential in metrological applications especially on rough surfaces,
- the detailed electronic structure of the sample and of the tip (as an interacting system), which should be known especially in case of the high-resolution scanning tunneling spectroscopy.

In the following pages we will demonstrate these problems using two types of examples:

- surface profiles on a nanometer-size scale, and
- locally high-resolution scanning tunneling spectroscopy on organic adsorbates.

APPLICATION TO NANOMETER PROFILOMETRY

In connection with the development of a future nanotechnology, there occurs the problem of how to measure and calibrate nanometer-scale surface profiles. This problem is of special interest because of the interplay of classical and quantum mechanical aspects right on this scale.

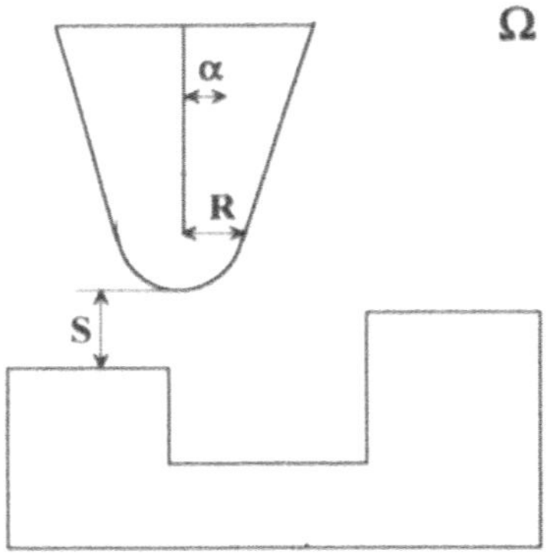

Figure 1. Model for the calculation of nanometer surface-profiles (schematic).

To study these problems in detail, we have developed a model seen in Figure 1. It is characterized by noninteracting electrons within arbitrarily shaped square-well potentials which mimic, in our case, the tip and sample electrodes, respectively. The solution of Schrödinger's equation is carried out numerically on a fine artificial mesh extending over the whole volume Ω. On the surface Ω of the latter, there has to be fulfilled the boundary condition:

$$\psi(\vec{r}) = 0 \text{ for } \vec{r} \in \partial\Omega .$$

(For more details the reader is referred to reference 4.) The advantages of such a model are:

- the electronic structures of tip and sample are treated on the same level;
- it enables a nonperturbative treatment of the tip-sample interaction;
- it allows a simple inclusion of quite complex geometrical configurations.

On the other hand, the drawbacks of the same model are:

- the restriction to noninteracting electrons, i.e., the neglect of screening;
- the neglect of the atomic structure of both electrodes (which can be overcome in principle);
- it is therefore not specific to the actual electrode materials;
- the extension to large three-dimensional systems is restricted for reasons of computer memory.

Figure 2 demonstrates some examples. (Tip and sample electrodes are square-well potentials, 8 eV deep, and the tunneling states are about 4 eV above the bottom. The tips are always assumed to have flank angles of 90°.) Figure 2(a) shows contour lines of constant LDOS above of a groove in the sample surface which is 2 nm broad and 1 nm deep. This is the STM line scan profile as expected according to the standard model.[2] The structure of a one-electron state of a tip electrode is visualized in Figure 2(b) by the probability-density distribution $|\psi_i^{(1)}|^2$ of that state.

Figures 2(c) and (d) show constant current profiles calculated for the same groove as in Figure 2(a) using tips with radii of curvature R = 8 Å and 20 Å, respectively. Clearly, the geometrical restrictions of real 3-D imaging due to the tip shape can be seen: with increasing tip radius resolution becomes worse. On the other hand, these images may be deconvoluted to some extent according to reference 5 in case of known tip shape (e.g. from an electron micrograph). This procedure is unique as long as there is no coincidence of the tangential slopes of the tip and sample electrode. These points of coincidence are marked in Figures 2(c) and (d), below. As a result, at least the lateral positions of the groove edges can be recovered quite accurately. The vertical shape of the profile can at least partially be obtained from the scaling behavior of the coincidence points with tip-sample distance.[4]

In the experimental example shown in Figure 3, an IC-structure is imaged by AFM using two different tips (in this case only the classical geometrical aspects of the previous discussion are relevant). The one tip is of the standard pyramidal shape and the other one is a supertip grown on a standard tip by electron beam deposition in a scanning electron microscope (Figure 3(b)). The cross-sections along the line marked in Figure 3(a) and shown in Figures 3(c) and (d) obviously demonstrate the better vertical resolution of the supertip due to decreased radius of curvature at the apex and due to the dramatically diminished flank angles.

APPLICATION TO SINGLE ORGANIC MOLECULES

Adsorbed organic molecules (of relatively large size) are often quite easy to investigate by STM even at ambient conditions.

Figure 4 shows as an example ordered (self-assembled) layers of an oxy-cyano-biphenyl-alkane (5OCB - the number 5 refers to the number of C atoms in the alkane chain) on highly oriented pyrolytic graphite (HOPG). This substance is assembled by scanning the tunneling tip over the surface area, which is then imaged in the following raster scans.[6] Figure 4(b) is a constant height image of such an ordered region including a domain boundary to another differently oriented area. Figure 4(a) conforms to the chemical structure of the corresponding 5OCB molecules, and Figure 4(c) is a structure proposed for that layer. We have found in this case remarkable differences with respect to the well-known cyano-biphenyl-alkanes (CB's).[6,7] It should even be possible to study phase transitions within such systems by direct visualization.

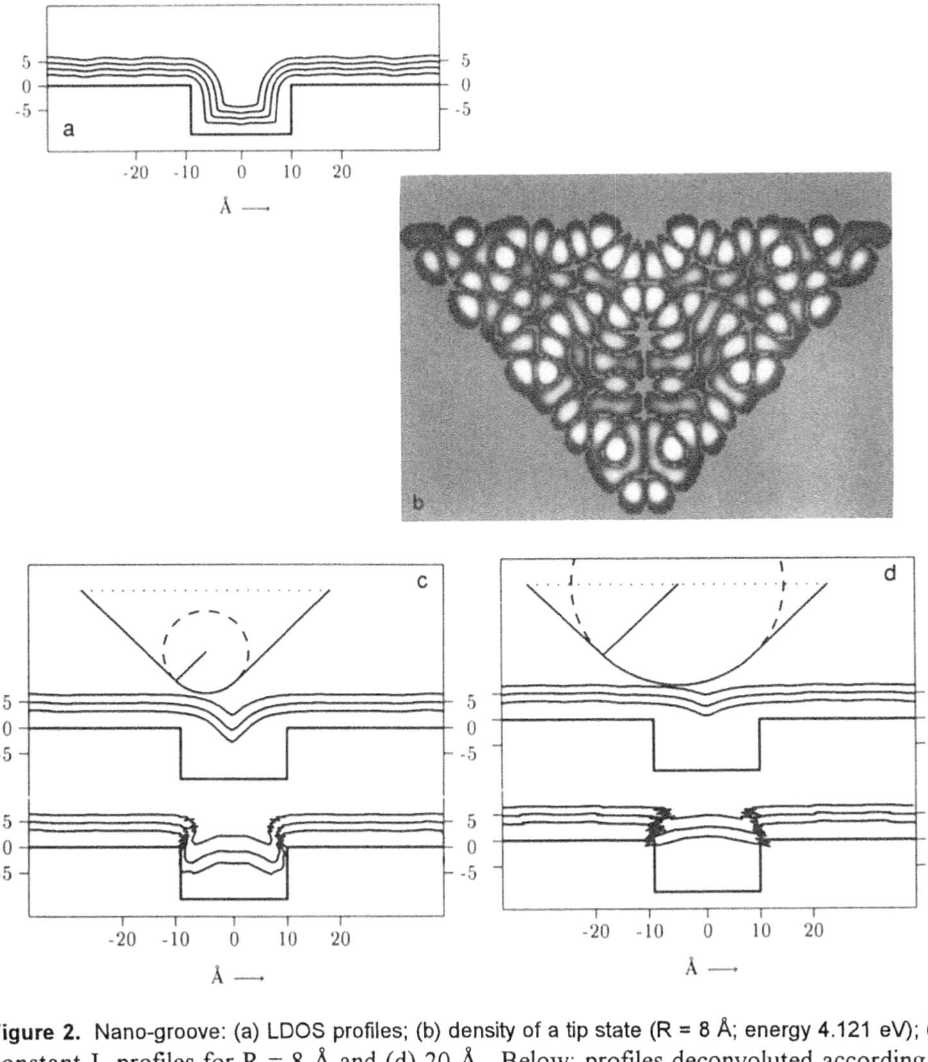

Figure 2. Nano-groove: (a) LDOS profiles; (b) density of a tip state (R = 8 Å; energy 4.121 eV); (c) constant I_T profiles for R = 8 Å and (d) 20 Å. Below: profiles deconvoluted according to reference 5.

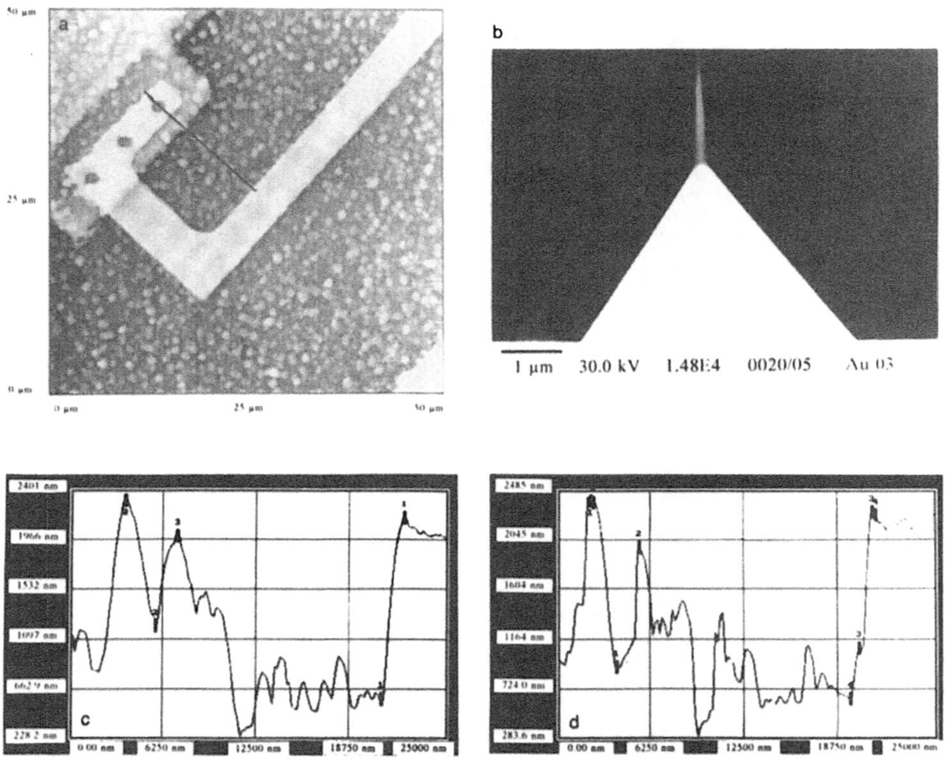

Figure 3. Influence of the geometrical tip shape in topographic imaging of (a) an IC-structure (c) without and (d) with an AFM supertip (seen in (b)).

Actually, this type of structural investigations offers some submolecular structure which is similar to the total charge density rather to the corresponding LDOS. This is due to the rather fuzzy imaging conditions at ambient atmosphere.

For a more detailed local electronic spectroscopy which is possible with STM an ultra-high vacuum (UHV) environment is essential. Since corresponding systematical investigations are still on the way, we present here some theoretical calculations of the LDOS of isolated molecules.[8]

The local orbital density (which corresponds to the LDOS in the case of a single molecule) may be obtained by solving the Hartree-Fock equations:

$$- \nabla^2/2 \; \chi_i(\vec{x}) + V_{ext}(\vec{r}) \; \chi_i(\vec{x}) +$$

$$+ \sum_j \int d\vec{x}' \, |\chi_j(\vec{x}')|^2 / |\vec{r} - \vec{r}'| \; \chi_i(\vec{x}) -$$

$$- \sum_j \int d\vec{x}' \; \chi_j^*(\vec{x}')\chi_i^*(\vec{x}') / |\vec{r} - \vec{r}'| \; \chi_j(\vec{x}) = \epsilon_i \, \chi_i(\vec{x})$$

$$\vec{x} = \{\vec{r}, s_z\}.$$

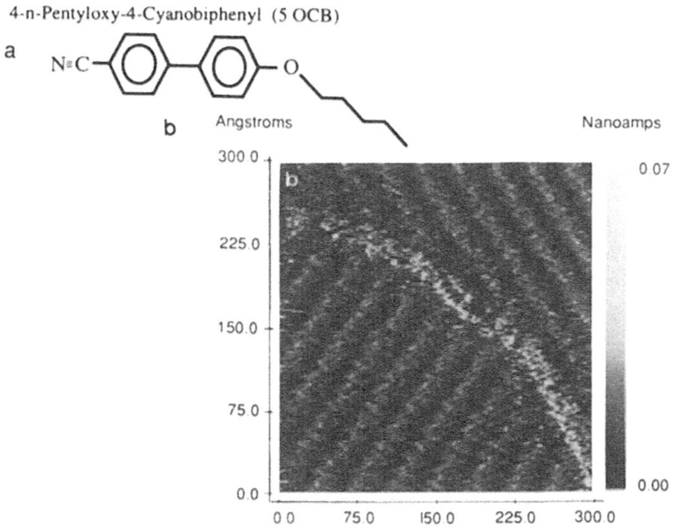

Figure 4. 5OCB (a) and STM constant-height image of a self-assembled layer of that substance on HOPG (b).

The summations are over all occupied molecular orbitals, and V_{ext} is the electrostatic potential due to the nuclear skeleton of the molecule.

Figure 4(d) shows as a result the total electron density of a single 5OCB molecule obtained by summing up all LDOS's of the occupied orbitals.[8]

Based on such type of calculations for (oxy-)cyano-biphenyl-alkanes and for metal-phthalocyanines, we could predict topological effects in spectroscopic images not only due to the imaging of different orbitals (by changing U_T) but also due to the imaging of different iso-LDOS surfaces of the same orbital (by changing I_T). These results will be published elsewhere.[8,9]

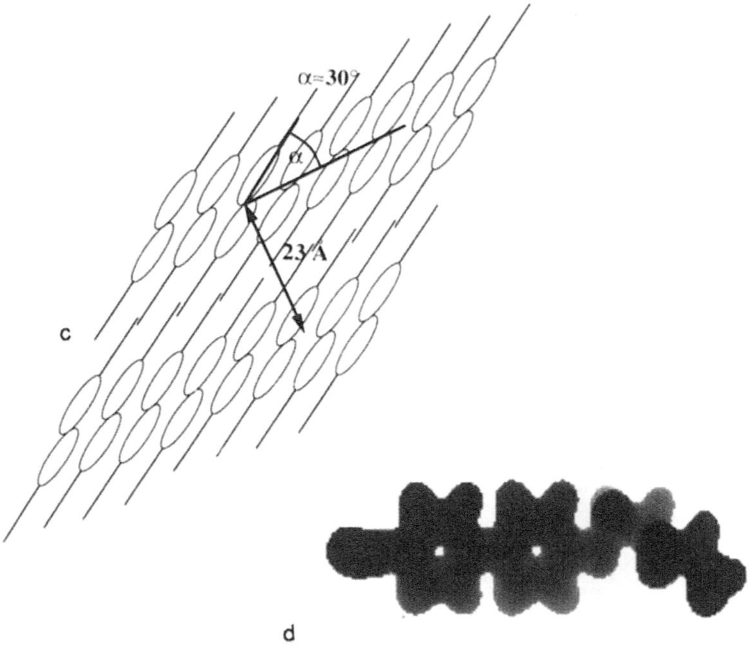

Figure 4 (Continued). Proposed structure of the ordered molecular layer of 5OCB (c) and calculated total electron density of a single molecule (d).

Using an approach as outlined in this section, the complex effects of adsorbate-substrate and tip-adsorbate interaction have not yet been included.

SUMMARY

We have demonstrated by some examples that the necessary quantification in future scanning probe microscopy demands strong interaction between theory and experiment. We believe that the ultimate power of SXM can be further developed only in this way.

REFERENCES

1. M. Hietschold, O. Pester, W. Vollmann, A. Heilmann, P. Stäbeler, H. Sbosny, X. Grählert, H.-U.Sonntag, A. Bruska, B.Winzer, T. Schimmel, and L. Koenders, STM and AFM investigations on organic-material thin-films and adsorbate particles in air, *in:* "Atomic Force Microscopy/Scanning Tunneling Microscopy," Samuel H. Cohen, Mona T. Bray, and Marcia L. Lightbody, eds., Plenum Press, NY, 43-52 (1994).
2. J. Tersoff, D.R. Hamann, Theory of the scanning tunneling microscope, *Phys.Rev.* B31, 805 (1985).
3. J. Bardeen, Tunnelling from a many-particle point of view, *Phys.Rev.Lett.* 6, 67 (1961).
4. H. Sbosny, L. Koenders, M. Hietschold, Calculation of STM profiles for the nanometrology, *Thin Solid Films* 264, 273 (1995).
5. G. Reiss, F. Schneider, J.Vancea, H. Hoffmann. Scanning tunneling microscopy on rough surfaces: deconvolution of constant current images, *Appl.Phys.Lett.* 57, 867 (1990).
6. M. Hietschold, K. Walzer, D. Porezag, Scanning tunneling microscopy on liquid-crystal films deposited on layered material surfaces, *Scanning* 17, V53 (1995); also M. Hietschold, K. Walzer, Molecular imaging of ordered liquid crystal structures on layered materials by STM at ambient conditions, *J. Vac. Sci Technol.* B. (in press).
7. D.P.E. Smith, J.K.H. Hörber, G. Binnig, H. Nejoh, Structure registry and imaging mechanism of alkylcyanobiphenyl molecules by tunneling microscopy, *Nature* 344, 641 (1990).
8. M. Hietschold, D. Porezag, G. Wolf, Calculation of molecular-orbitals of metal-phthalo-cyanines and (oxy-)cyano-biphenyl-alkanes in relation to scanning tunneling spectroscopy, to be published.
9. O. Pester, D. Porezag, M. Hietschold, Molecular-resolution images of copper- and lead-phthalocyanine single crystals by atomic-force microscopy, submitted for publication.

OBSERVING REACTIONS VIA FLOW INJECTION SCANNING TUNNELING MICROSCOPY

James D. Noll, Paul G.Van Patten, and M.L. Myrick
Department of Chemistry and Biochemistry
University of South Carolina
Columbia SC 29208

Abstract: The scanning tunneling microscope (STM) is used to view conductive and semiconductive surfaces to obtain topographic and structural information. A number of literature reports describe molecules absorbed on surfaces. We describe a flow injection system which allows imaging before, during, and after a surface reaction. The flow injection system consists of a flow cell in which a solution is pumped over a sample via a flow injector and a peristaltic pump during STM imaging. Results indicate that atomic imaging can be maintained under a rapidly-flowing solution stream. This system can provide a way to observe reactions occurring on surfaces. Preliminary applications of the system that include the etching of a metal surface, attachment of thiols on a gold surface, and attachment of polymers onto highly ordered pyrolytic graphite (HOPG) step defects are described.

INTRODUCTION

The scanning tunneling microscope (STM) can be used to view the surfaces of a variety of conductive and semiconductive materials such as highly ordered pyrolytic graphite and gold, giving information about structure and surface topography. The observation of chemical species bound to surfaces has also received much attention; HOPG and gold have become common substrates for such molecular absorption studies.[1-6] However, defects can be easily mistaken for molecules on the surface especially if the size or structure of the defect matches that expected for the reactant. For example, Beebe[7] and Myrick[8] reported defects with many of the same characteristics reported for biological molecules adsorbed on HOPG. The characterization of molecules absorbed on gold is more certain because some molecules (e.g. thiols) are known to bind strongly to gold, unlike HOPG. However, even in this case the confirmation of the presence of molecules is based on observation of surface morphologies that are consistent with expectations, since atomic resolution is rarely obtained.

Methods to confirm the binding of a molecule to a surface are needed. A flow injection system would be a step toward solving this problem since images of the same area before,

Atomic Force Microscopy/Scanning Tunneling Microscopy 2
Edited by S.H. Cohen and M.L. Lightbody, Plenum Press, New York, 1997

during, and after the binding reaction could be obtained to confirm that observed structures are not substrate-only features.

An additional use of a flowing solution reactor stage for STM would be to control the growth of molecules on a substrate. Molecular layer deposition has been used to control the growth of polymers on a substrate,[9] but this method is limited to one-dimensional growth control. With a flow injection system, reagents that anchor to activated sites on surfaces could be introduced to the substrate while imaging, and reagents could be varied in type and concentration to produce surface-attached structures.

In this paper we briefly discuss the construction and evaluation of the flow injection system. We then show some preliminary data of reactions on a surface to demonstrate the power of this system.

EXPERIMENTAL

The overall design of the flow injection system (see Figure 1(a),(b)) consists of a flow cell that fits into the STM base of the Digital Instruments Nanoscope II where fluids arepumped in and out via a peristaltic pump and a Rheodyne HPLC injector. The solvent is pumped through the injector, where different chemicals can be inserted into the flowing solvent, to the flow cell.

Before each experiment, the 80/20 PtIr cut tips were coated with polyethylene glue from a glue gun to eliminate the faradaic and nonfaradaic currents present in the solution (in this case, the distilled water or ethanol). The tip was completely coated with the glue and then heated with a heat gun so that the excess glue ran off the tip.

The Manostat peristaltic pump was then adjusted so that distilled water was pumped in and out at comparable rates so that the cell did not overflow or become dry.

After the cell was emptied, the cell and STM head were placed on the base, and the tip was adjusted closer to the surface. Atomic resolution of HOPG was then obtained, and the tip was withdrawn. The liquid was then placed in the cell, the tip was reengaged, and the flow was initiated. The scan size was set at 0 nm (to immobilize the tip in the x and y directions) in constant height mode and the z-output was measured via a Nicolet Pro 50 digital storage oscilloscope. In constant height mode at zero scan size, the z-output should be a horizontal line. Any deviation from this is "noise" or scanner error. The noise was measured at a variety of flow rates, gain settings, and with several coated tips. Fourier transforms for each z-noise recording were then performed and averaged to study additional noise in the z-position that the fluid flow causes.

After the tips were used, each tip was just placed in the water solution without being engaged, and the head offset was recorded to indicate the magnitude of faradaic currents resulting from imperfect insulation of the tip. A scanning electron microscope image was obtained for each coated tip to view the quality of the coating of the tip.

Distilled water was replaced with ethanol, and the injector was inserted in series with the flowing solvent (ethanol). Freshly cleaved mica was heated for one hour at about 673 K. Then the mica was sputtered with gold while it was still hot and annealed at about 673 K for about 16 hours. The gold-coated mica was placed in the flow cell in the STM, and 30 % hydrogen peroxide was injected into the injector and allowed to flow at a rate of 0.6 mL/min with ethanol to the cell. The setpoint current was measured as a function of time.

Ethanol was flowed at a rate of 1.2 mL/min over the gold-coated mica. 30% hydrogen peroxide was injected into the flow and timed. Immediately after injection, an image of the gold was obtained. After one minute, several images were captured until two minutes had passed and the images ceased changing. The gold was not completely clean. A saturated solution of l-octadecanethiol was then injected into the injector and a series of images was captured to observed the reaction of the thiols on the gold surface.

Gold-coated mica was annealed for 15 hours and then set in a peroxide solution (0.8 mL 30% hydrogen peroxide, 5.0 mL concentrated NH40H, and 29.0 mL distilled H2O) for 2 minutes.

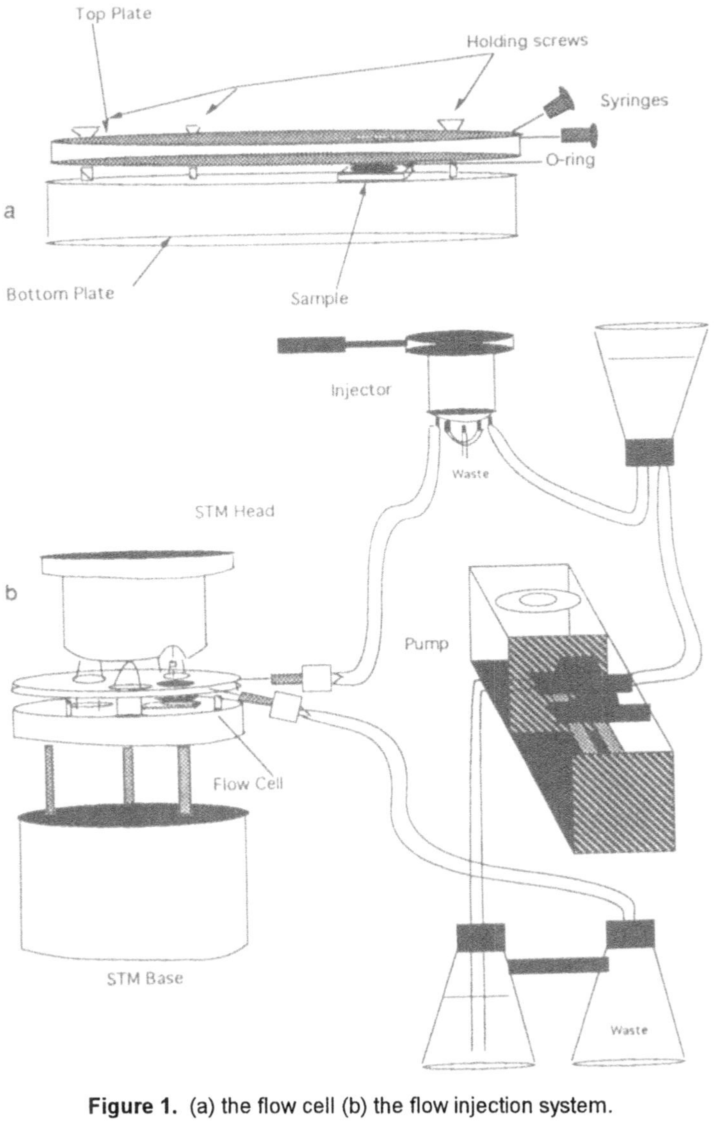

Figure 1. (a) the flow cell (b) the flow injection system.

This provided a clean gold surface. The mica was placed under the STM and a saturated solution of l-octadecanethiol in ethanol was again injected into the system. A clean gold-coated mica sample was made as described previously, and a l mM solution of the l-octadecanethiol in ethanol was produced. The FISTM experiment above was repeated using the new sample and diluted solution. The saturated and diluted 1octadecanthiol solutions were placed into a Beckman System Gold high performance liquid chromatography with a size exclusion column (Beckman SEC 3000) and 100% ethanol as the eluent.

Drops of hot solution of concentrated nitric and sulfuric acid was placed onto a highly ordered pyrolytic graphite (HOPG) monochromator. The monochromator was rinsed with distilled water and drops of a hot saturated SnCl in conconcentrated HCl solution was added. After the SnCl solution was rinsed, a hot solution of NaOH was added. The graphite was observed under the STM, and a good step defect was searched for and concentrated on. First pyromellitic dianhydride (PMDA) and then 2 minutes later 4,4'-diaminodiphenyl ether (DDE) were injected into the sytem. A series of images were captured that watched the attachment of the polymers onto the step defect.

RESULTS AND DISCUSSION

Noise

No additional noise in the z-component resulted from the flow of the liquid. The flow rates and the gains also had no effect on the noise picked up on the oscilloscope. However, the peristaltic pump did increase the noise. The reason for this is not known yet, but decoupling the pump did decrease its effect. Drift did not seem to cause a problem, and atomic resolution was even obtained (see Figure 2). From these results we can see that the flow injection scanning tunneling microscopy is possible. Now we want to look at different applications of this system.

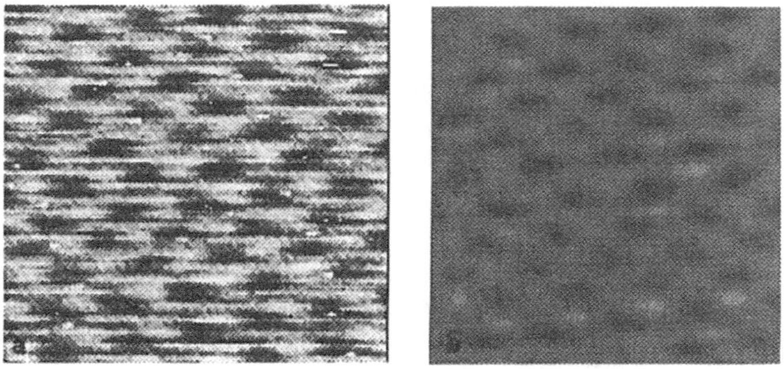

Figure 2. Atomic resolution of HOPG with a flow of distilled water at 0.6 mL/min (a) unfiltered (white dots are noise from the peristaltic pump) (b) filtered.

Head Offset

The head offset includes a measure of non-tunneling current passing through the tip. When the offset exceeds the setpoint current, the z-center goes out of range and the tip will engage immediately. Hydrogen peroxide added to the fluid stream causes the head offset to increase during its residence in the cell. If the setpoint current is not initially high enough, the tip retracts out of range when the offset increases; therefore, the head offset was measured as a function of time to determine that it will maintain tunneling contact, the minimum setpoint current, and the time of passage of the peroxide over the sample. Figure 3 shows a graph of head offset vs. time. The peroxide took approximately 90 seconds to enter the cell after injection with a flow rate of approximately 0.6 mL/min. and remained in the cell for about 30 seconds. The highest setpoint current measured was 3 nA. Therefore, if the setpoint current is set above 3 nA, the tip will remain in tunneling range on the same area even after the peroxide enters the cell.

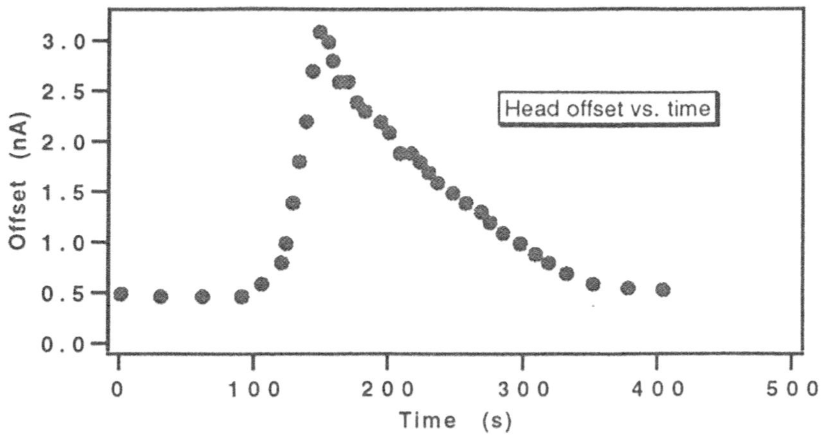

Figure 3. Graph of head offset vs. time.

Etching

In this laboratory, hydrogen peroxide is used for etching gold and cleaning from its surface contaminants such as thiols. The flow injection system was used to observe this etching. Figure 4(a) shows the surface of gold immediately after the 30% hydrogen peroxide was injected into the flow of ethanol before it reached the flow cell. Figure 4(b)-(c) shows the same gold surface after the peroxide solution was washed over the sample. The images shown in Figure 4 are from a sequence of images acquired over a period of one minute showing the step-by-step progress of the etching reaction. As Figure 4 shows, a pit located in the upper middle of the Figure 4(a) is rapidly attacked by the peroxide and enlarged in an anisotropic fashion.

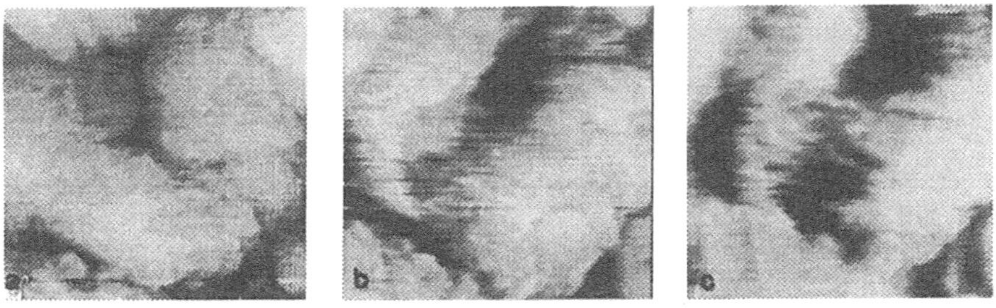

Figure 4. STM images of hydrogen peroxide flowing over gold. (a) is an image of the gold surface right after the injection of 30% hydrogen peroxide before it entered the flow cell. (b) is an image of the surface as the peroxide is entering the cell, and (c) is after the peroxide was washed out of the cell.

Thiol Attachment

Thiols have been routinely deposited on atomically flat gold samples and then imaged under the STM.[10-12] The thiols form well-ordered monolayers and atomic resolution has been achieved in the imaging of these thin films.

We wanted to see this attachment of thiols as it occurred. 1-octadecanethiol was injected into the FISTM and observed. The thiols tore large clumps of the surface (see Figure 5). This must be due to the thiols tearing contamination on the gold surface.

When the thiols were attached to the clean gold surface (see Figure 6), large clumps were formed and part of the thiols were washed away leaving a film on the surface. This did not leave a well-ordered monolayer as we expected. We believe that this could be because aggregates are being formed in the solution in the form of micelles. These aggregates would form strong bonds with the gold. After they attach to the gold, the micelles break up leaving a thin layer on the gold surface.

Further evidence of the micelle model is when we tried a diluted solution of the octadecanthiol (lmM) on clean gold, we did not see the formation of the clumps. This concentration must be too low to form aggregates. When both the saturated and diluted solutions were placed into the HPLC (see Figure 7) with size occlusion column, the diluted solution showed a large peak at 11.8, and the saturated solution showed a large peak at 11.2 and a shoulder at 11.8. The peak at 11.2 min. is the result of the aggregates in the saturated solution.

Polymer Attachment

In a previous study, HOPG step defects seemed to be a result of grain boundaries formed during the manufacturing of the HOPG monochromator.[13] Since these step defects are more reactive than the basal plane, we wanted to attach polymers to these defects in order to control the polymer growth. The polymer would grow along the step and be limited by the size of the step defect.

Yoshimura et al.[14] used PMDA and DDE to demonstrate molecular layer deposition. The epoxy group and amine group of the two polymer will attach to one another but will not attach to itself.

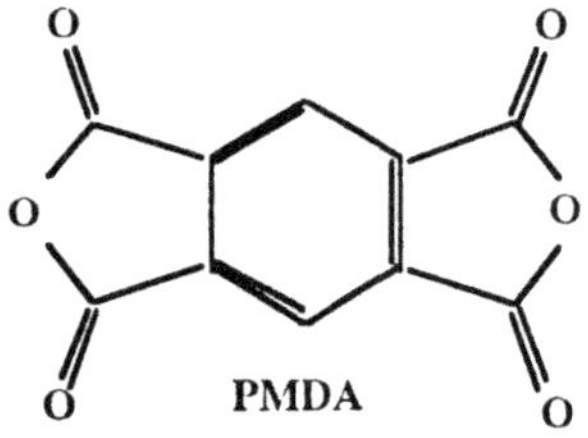

PMDA

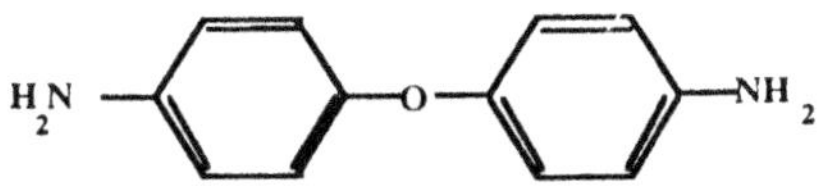

DDE

We want to try to step these polymers onto a HOPG step defect. First, the step defect has to be derivatized so that one of the polymers will attach to the ledge. Since the step is not part of the basal plane, some functional group has to form where the basal plane terminates. An aromatic group could form at these grain boundaries. We derivatized the step with the following reactions:

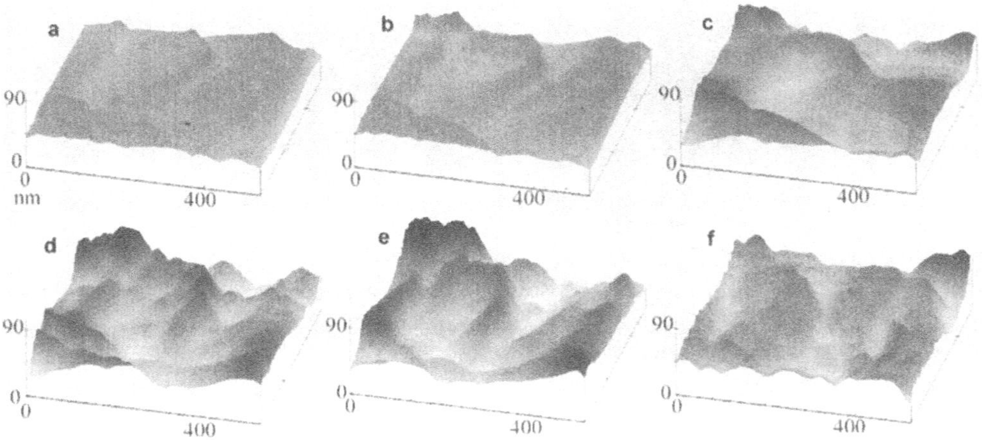

Figure 5. A series of STM images of 1-octadecanethiol attaching to a dirty gold surface. (a) is the gold surface before the thiols hit; (b) is right when the thiols hit; (c) - (e) are images in series; (f) is about 10 minutes after the thiols hit.

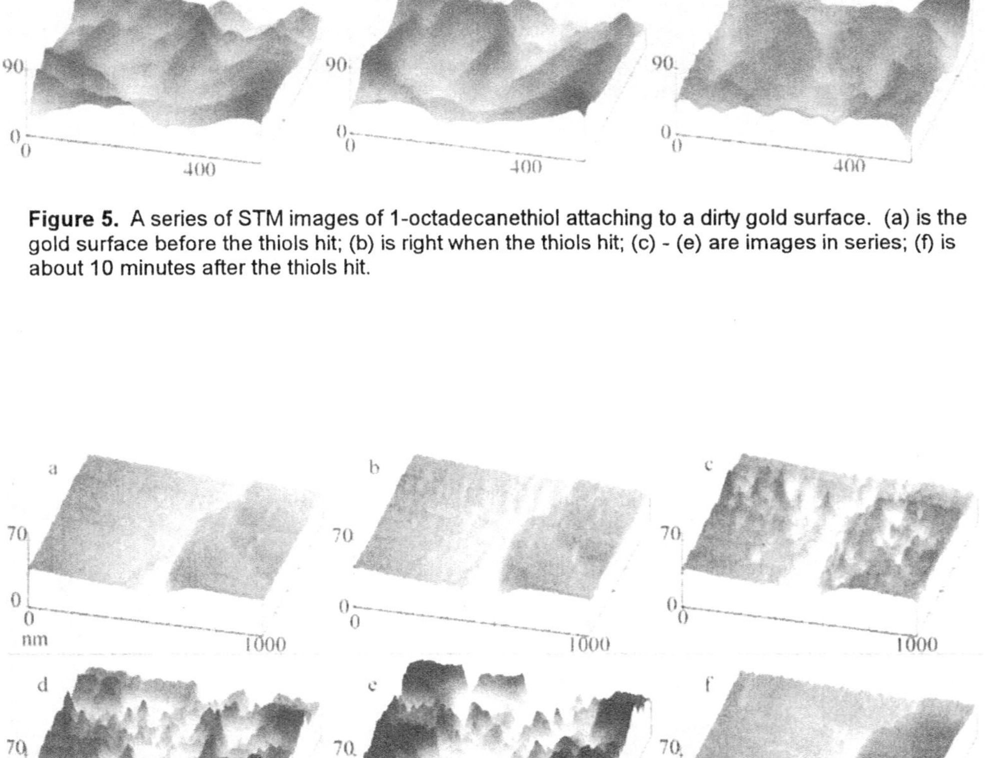

Figure 6. A series of STM images of 1-octadecanethiol attaching to a clean gold surface. (a) is the gold surface before the thiols hit; (b) is right when the thiols hit; (c)-(e) are images in series; (f) is about 10 minutes after the thiols hit.

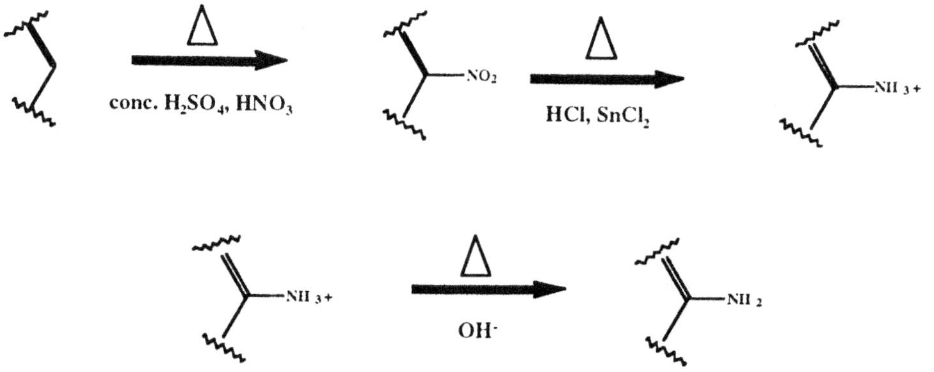

This formed an amine group onto the ledge, and the PMDA will attach to the amine group.

The PMDA was then injected into the FISTM and images were captured. This experiment was done three times, and each time the attachment of the polymers was visible (see Figures 8, 9). The experiment was also performed without derivatizing the ledges, and no change was observed on the surface.

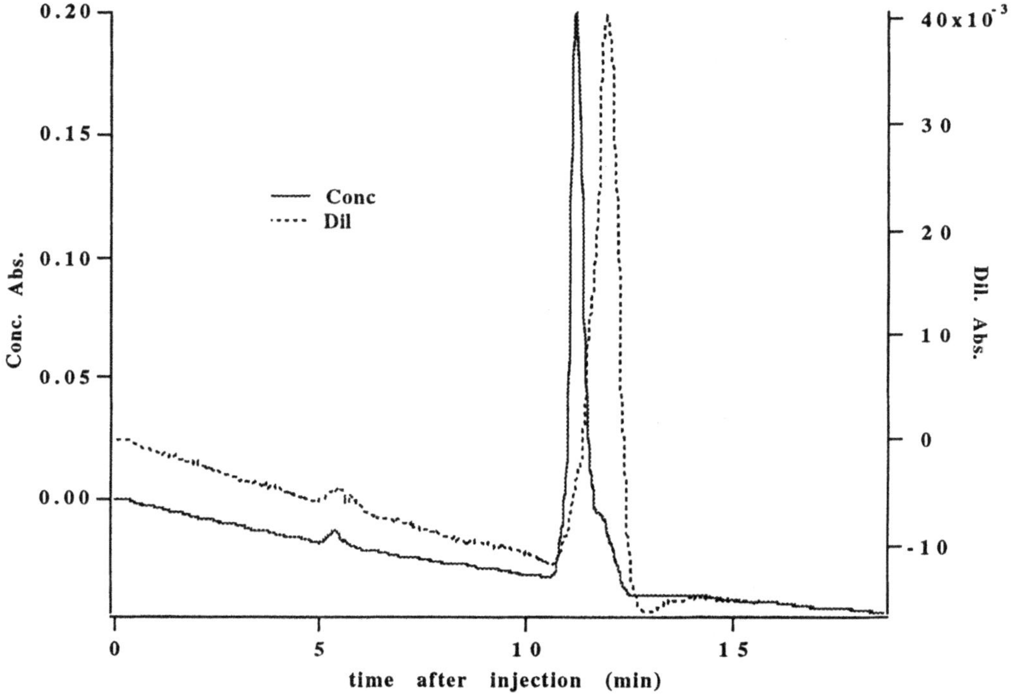

Figure 7. HPLC plot of the saturated and diluted 1-octadecanethiol solutions. The dotted line is the diluted and the solid line is the saturated. The diluted line has a large peak at 11.8 min and saturated line has a large peak at 11.2 and a shoulder at 11.8.

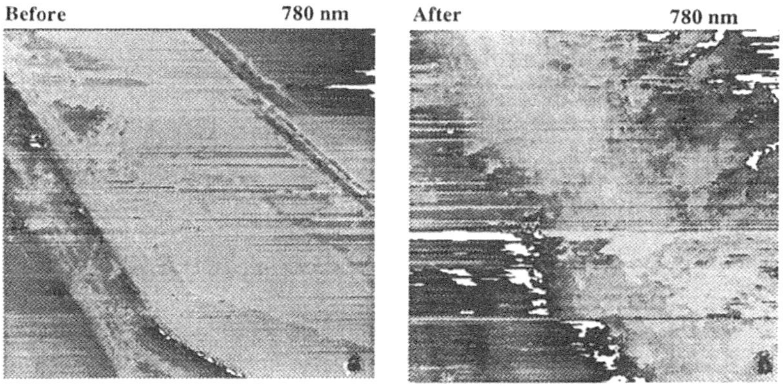

Figure 8. (a) is an STM image of a step defect on HOPG. (b) is the same step defect but after the surface has been exposed via FISTM to PMDA and DDE 10 times.

Figure 9. (a) An HOPG step after the surface has been exposed to PMDA and DDE twice; (b) after 4 steps of PMDA and DDE; (c) after 10 steps.

CONCLUSION

Flow injection scanning tunnelling microscopy (FISTM) adds a dimension to the STM where reactions can be observed while they occur. Cleaning of gold and the attachment of thiols and polymers have been demonstrated. Since the FISTM gives us the capability to obtain a before and after picture, we are able to observe these reactions.

ACKNOWLEDGMENTS

The authors thank Dr. Norbert Bikales and the National Science Foundation (Grant No.: DMR-9207460) for support of this work. The authors also appreciated the useful discussions with Dr. Yunke Zhang.

REFERENCES

1. J.S. Foster, J.E. Frommer, and P.C. Arnett, Monecular manipulation using a tunnelling microscope, *Nature*, 331:324-326 (1988).
2. Y.L. Lyubchenko, S.M. Lindsay, J.A. DeRose, and T. Thundat, A technique for stable adhesion of DNA to a modified graphite surface for imaging by scanning tunneling microscopy" *J. Vac. Sci. Technol.* A9:1288-1290 (1991).
3. M. G. Youngquist, R. J. Driscoll, T.R. Coley, W.A. Goddard, and J.D. Baldeschwieler, Scanning tunneling microscopy of DNA atom-resolved imaging general observation and possible contrast mechanism, *J. Yac. Sci. Technol.*, A9:1304-1308 (1991).
4. W. Li, J.A. Virtanen, and R.M. Penner, A nanometer-scale galvanic cell, *J. Phys. Chem.*, 96:6529-6532 (1992).
5. J. F. Womelsdorf, W. C. Ermler, and C. J. Sandroff, Imaging of colloidal gold on graphite by scanning tunneling microscopy: isolated particles, aggregates, and ordered arrays, *J. Phys. Chem.*, 95:503-505 (1991).
6. T. J. McMaster, H. Carr, M. J. Miles, P. Cairns, and V. J. Morris, Polypeptide structures imaged by the scanning tunneling microscope, *J. Vac. Sci. Technol.*, A8:648-651(1990).
7. C. R. Clemmer and T. P. Beebe, Graphite: A mimic for DNA and other biomolecules in scanning tunneling microscope studies, *Science*, 251:640-642 (1991).
8. M. L. Myrick, N.V. Hud, S. M. Angel, and D.G. Garvis, *Chemical Physics Letters*, 180:156-160 (1991).
9. T. Yoshimura, S. Tatsuura, and W. Sotoyama, Polymer films formed with monolayers growth steps by molecular layer deposition, *Appl. Phys. Lett.*, 59:482-484 (1991).
10. Y.T. Kim and A. J. Bard, Imaging and etching of self0-assembled n-Octadecanethiol layers on gold with the scanning tunneling microscope," *Langmuir*, 8:1096-1102 (1992).
11. C.A. Widrig, C.A. Alves, M.D. Porter, Scanning tunneling microscopy of ethanethiolate and n-Octadecanethiolate Monolayers spontaneously abserbed at gold surfaces, *J. Anl. Chem. Soc.* 113:2805-2810 (1991).
12. N. J. Tao and S.M. Linsay, Observations of the 22x30.5 reconstruction of Au(111) under aqueous solutions using scanning tunnelling microscope," *J. Appl. Phys.* 70:5141-5143 (1991).
13. J.D. Noll. J.B. Cooper, and M.L. Myrick, Analysis of highly ordered pyrolytic graphite step defects via scanning tunneling microscopy, *J. Vac. Sci. Tech.* B, B11:2006-2011 (1993).
14. T. Yoshimura, S. Tatsuura, and W. Sotoyama, Polymer films formed with monolayers gorwth steps by molecular layer deposition, *Appl. Phys. Lett.*, 59:482-484 (1991).

ADVANCES IN PIEZORESISTIVE CANTILEVERS FOR ATOMIC FORCE MICROSCOPY

M. Tortonese
Park Scientific Instruments
1171 Borregas Ave.
Sunnyvale, Ca 94089

Abstract: Piezoresistive cantilevers offer a novel detection scheme for Atomic Force Microscopy (AFM) in which no optics and no alignments are required to measure the deflection of the cantilever. The cantilever deflection is measured through the resistance of a stress sensitive resistor–a piezoresistor–integrated in the silicon cantilever. The use of piezoresistive cantilevers simplifies the operation of the microscope, especially for applications in ultrahigh vacuum (UHV) and at low temperature, where other detection schemes are difficult to implement. This paper reviews the principle of operation of piezoresistive cantilevers and presents recent results obtained using piezoresistive cantilevers for imaging in air, in water, in ultra high vacuum, at low temperature, on magnetic samples, for lateral force microscopy, in contact and noncontact modes.

INTRODUCTION

An AFM works by measuring the deflection of a cantilever beam which is raster scanned across a surface. The cantilever has a tip which can be in contact with the sample, or can be flying a short distance above the sample. In the latter case the cantilever is usually vibrated at or near its resonant frequency. No matter what the mode of operation is, one needs to measure the deflection of the cantilever. Several techniques have been implemented to do this. The first technique ever used was vacuum tunneling.[1] In this technique a metal tip is brought within tunneling range of a conductive cantilever and the deflection of the cantilever is measured through the variations in the tunneling current between the tip and the cantilever. This method requires subnanometer accuracy in the positioning of the tip with respect to the cantilever. In addition, the tunneling current is affected by contamination on the back of the cantilever. For these reasons this method has been almost completely abandoned.

In the interferometric detection method[2] an optical cavity is formed between the end of a fiber and the back of the cantilever. The interference between the light reflected off the end of the fiber and the light reflected off the cantilever is used to measure the cantilever deflection.

Atomic Force Microscopy/Scanning Tunneling Microscopy 2
Edited by S.H. Cohen and M.L. Lightbody, Plenum Press, New York, 1997

This technique is capable of measuring a deflection of 0.01 $Å_{rms}$ in a bandwidth from 100 Hz to 1 kHz[3] and is typically used when the highest sensitivity is required.

In the capacitive detection scheme,[4] the cantilever constitutes one of the two plates of a capacitor. The other plate is positioned parallel to the cantilever. The deflection of the cantilever alters the capacitance of the parallel plate capacitor. A minimum detectable deflection of 0.03 $Å_{rms}$ in the bandwidth from 0.1 Hz to 1 kHz has been demonstrated.[5]

The most common technique to measure cantilever deflection is the optical deflection,[6] in which the deflection of the cantilever is measured by measuring the position of a laser beam reflected off the back of the cantilever. Cantilever deflections of 0.1 $Å_{rms}$ in the bandwidth from 1 Hz to 1 kHz are measurable.[7] This method requires a laser to be precisely focused to the end of the cantilever and a photodetector to be aligned to the reflected beam.

All the techniques described so far require some external elements which need to be aligned to the cantilever in order to measure its deflection. Given the miniature size of the cantilever, the space available for the detector is typically limited. Consequently, in most cases the mechanics of the detector dominate the complexity of the instrument.

To simplify the construction and use of an AFM attempts have been made to build miniature[8,9] or integrated detectors.

One type of integrated detector is a piezoelectric cantilever.[10,11] The cantilever is fabricated from a piezoelectric material sandwiched between two electrodes. The strain in the cantilever caused by its deflection generates an electric field in the piezoelectric material and produces electric charges on the electrodes. By measuring the charge one can infer the deflection of the cantilever. Unfortunately the process of measuring the charge depletes the charge on the electrodes. No dc response is obtained from a piezoelectric cantilever. Therefore this technique has only been used when the cantilever is oscillated.

PIEZORESISTIVE DETECTION: PRINCIPLE OF OPERATION

In the piezoresistive cantilever the sensor is a silicon resistor integrated in a silicon cantilever. Silicon is a piezoresistive material.[12] This means that its bulk resistivity depends on the mechanical stress in it. When the cantilever is stressed by deflection the resistance of the integrated resistor changes. Therefore the deflection of the cantilever can be measured by measuring the resistance of the integrated resistor. This is shown schematically in Figure 1. Figure 2 shows a typical layout and cross-section of a piezoresistive cantilever. The piezoresistor runs across the cantilever from one contact, through the cantilever and to the

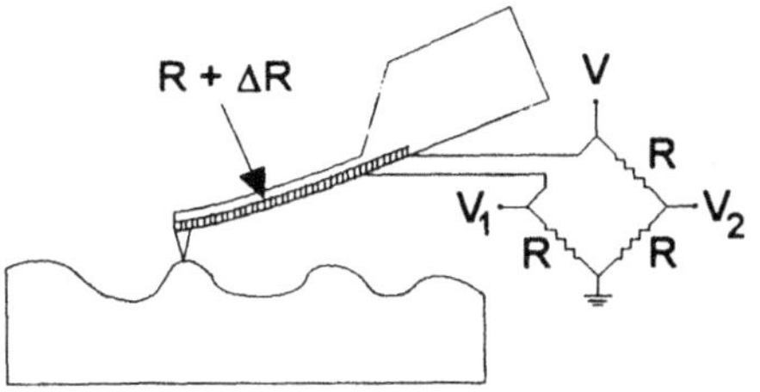

$$V_{out} = V_2 - V_1 = -V \frac{\Delta R}{4R}$$

Figure 1. Principle of operation of piezoresistive cantilevers.

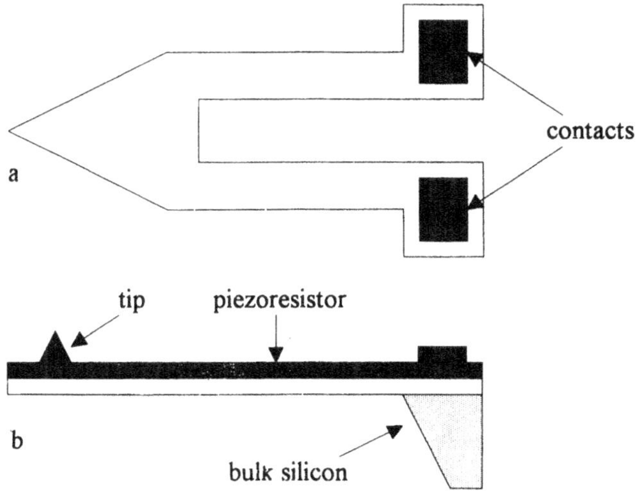

Figure 2. Plan view (a) and cross section (b) of a piezoresistive cantilever. The current flows from one contact to the other through the U-shaped cantilever. The piezoresistor is formed at the surface of the cantilever by doping.

other contact. There are two reasons why the resistance of the cantilever changes when it is deflected: the first is purely geometrical: as the cantilever is deflected, its top layer is stretched, while its width decreases. These two effects contribute an increase in resistance. This effect is called the geometric piezoresistance effect. The second reason is due to the fact that the stress in the silicon affects its band structure in a way that alters the population of electrons in bands with different value of electron mobility. This second effect is peculiar of semiconductors and in silicon is 100 times bigger than the geometric piezoresistance effect. Few materials have a higher piezoresistive coefficient than silicon, and none of them is amenable to conventional microfabrication techniques.[13]

Figure 3 shows an electron micrograph of a piezoresistive cantilever. The inset shows the silicon tip, which is integrally fabricated at the free end of the cantilever. The piezoresistive cantilevers that we have fabricated have resistances in the range of a few kΩ, they are a few hundred microns long, their thickness varies from 2 μm to 4 μm. Spring constants are in the range of 1 N/m to about 30 N/m. Resonant frequencies are in the range of 50 kHz to 200 kHz. The sensitivity, expressed in terms of fractional change in resistance per unit deflection is between 0.3×10^{-6} and 0.6×10^{-6} per Ångstrom of deflection. This means that when a cantilever with a resistance of 2 kΩ is deflected by 1 Ångstrom its resistance changes by 0.6 to 1.2 mΩ. Changes in resistance can be measured accurately with a resistive Wheatstone bridge. The minimum detectable deflection of the sensor is determined by the signal to noise ratio. The noise is ultimately limited by the thermal noise in the piezoresistor. In practice, with a supply voltage of a few volts, resistance changes of 1 mΩ can be detected in a 1 kHz bandwidth. This fact translates in a minimum detectable cantilever deflection of about 1 Å in a 1 kHz bandwidth. Given a typical spring constant of 10 N/m, this also translates into a minimum detectable force in the order of 10^{-9} N. Stiff cantilevers are more sensitive to deflection than soft cantilevers because a higher stress is required to produce the same deflection, and the piezoresistive cantilever essentially measures the stress in the cantilever.

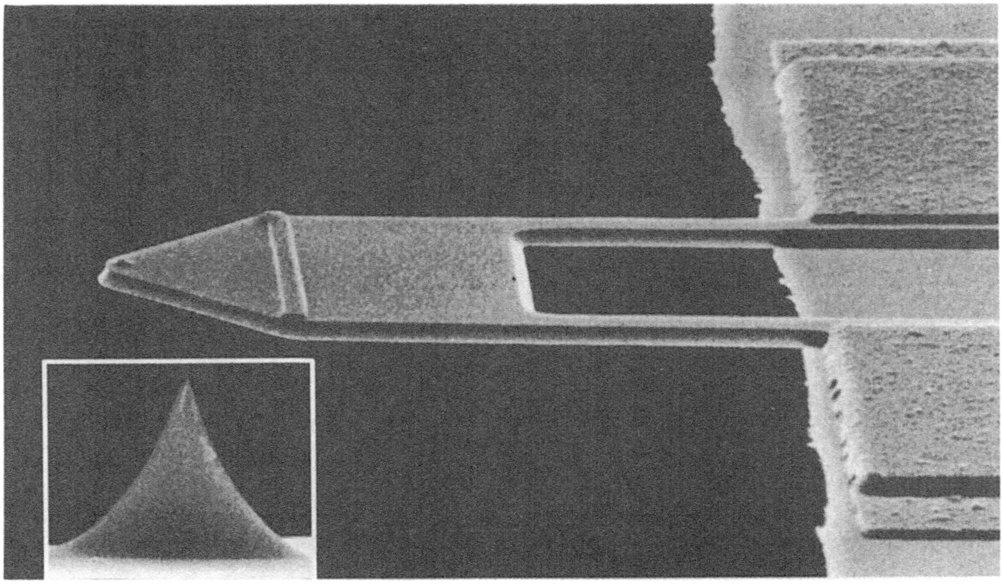

Figure 3. Scanning electron micrograph of a piezoresistive cantilever and tip (inset).

A minimum detectable deflection of 0.1 $Å_{rms}$ in the 10 Hz to 1 kHz bandwidth has been demonstrated with a cantilever having a spring constant of 100 N/m.[14] Typically a 2 kΩ cantilever is operated with a bridge supply of 2.5 volts. This means that the power dissipated in the cantilever is 0.78 mW, which is in the same order of the power of light absorbed by a cantilever when a detection system employing a laser is used. Silicon is an excellent heat conductor, so the heat generated within the piezoresistive cantilever is easily carried away to the silicon substrate. From experiments in vacuum it is known the increase in temperature in the piezoresistive cantilever at the conditions given above is only 2° C.[15]

IMAGING WITH PIEZORESISTIVE CANTILEVERS

Piezoresistive cantilevers with a spring constant in the range of a few tens of N/m are sensitive enough to image atomic corrugations on periodic crystals. Figure 4 shows atomic resolution on boron nitride - an insulator - obtained in contact mode, constant height mode, using a piezoresistive cantilever with a spring constant of 100 N/m.[14] This image does not really show the individual atoms. It rather represents the interaction of a large area on the sample with a large area on the tip, and can be explained with the theory of "phase-locked multitips".[16]

The contact mode of operation of an AFM has some drawbacks. Mainly the fact that the sharp tip, being in contact with the sample, has a tendency to scratch the sample or be damaged, unless hard and non-reactive samples are imaged.

New modes of operation have recently become more attractive. These include the non-contact[17] and the tapping[18] mode of operation. In these modes the tip is vibrated at or near its resonant frequency a distance away from the sample such that the tip never comes into contact with the sample (noncontact mode) or comes into contact only intermittently at each cycle

Figure 4. Image of boron nitride - an insulator - taken with a piezoresistive cantilever in contact mode showing the atomic lattice.

(tapping mode). For these applications stiff cantilevers are required, as soft cantilevers tend to get trapped into the sample. Typically cantilevers with spring constants larger than 10 N/m are used. This is the range of spring constant where piezoresistive cantilevers achieve a level of performance comparable, if not superior, to conventional optical deflection detection schemes.

The piezoresistive cantilever is particularly well suited for use in UHV because it does not require any alignment and is compatible with the baking process. Other detection schemes requiring alignment can be difficult to implement and operate in UHV. Figure 5 shows an image of a crystal of potassium chloride taken in non-contact mode in UHV with a piezoresistive cantilever.[15] The crystal was cleaved in air and then transferred into vacuum. The lateral resolution in this image is 20 Å, while the vertical resolution is such that individual monoatomic steps can be identified.

Piezoresistive cantilever can also be used in tunneling mode, as demonstrated by Giessibl. In this case the piezoresistor extends all the way to the tip of the cantilever. Consequently the cantilever can be biased through the contacts to the piezoresistor and the tunneling current is carried by the piezoresistive conductive channel. Figure 6 shows an image of Si(111) 7x7 collected in UHV in tunneling mode using a piezoresistive cantilever.[15]

Another important field of application for piezoresistive cantilevers is low temperature AFM. Low temperature AFM has traditionally been challenging because of the difficulties in building and aligning the cantilever deflection detector for low temperature operation. The piezoresistive cantilever eliminates this difficulty. In addition, piezoresistive cantilevers work better at low temperature than at room temperature for two reasons: the first reason is that the piezoresistive coefficient increases at low temperature.[19] Experimentally, the sensitivity of piezoresistive cantilevers increases by a factor of 2.4 going from room temperature to 6 K.[20] The second reason is that the thermal noise in the piezoresistor decreases at low temperature. Yuan et al. have used piezoresistive cantilevers to image VHS magnetic tape at low temperature. The piezoresistive cantilevers are coated with a thin layer of iron. Care must be exercised not to short the contacts to the piezoresistor. Images of magnetic domains on the VHS tape have been obtained down to a temperature of 6 K.[20]

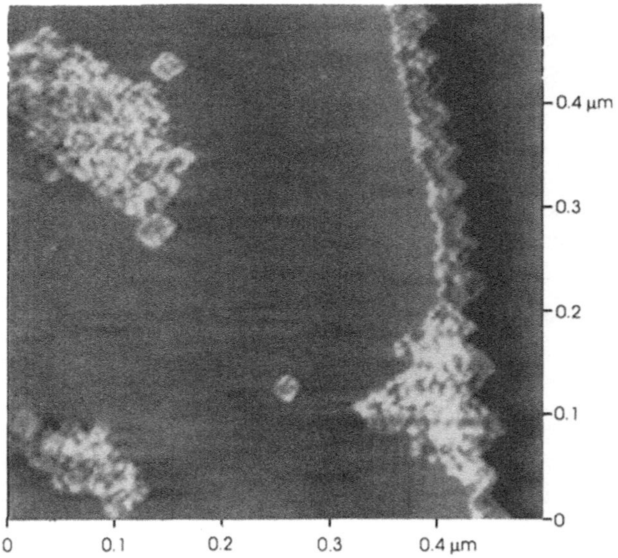

Figure 5. Image of potassium chloride taken in UHV with a piezoresistive cantilever in noncontact mode. A step edge along the [100] direction is visible. This edge is constituted of [110] facets. The lateral resolution is 20 Å and monoatomic steps are seen in the vertical direction.

Operation of piezoresistive cantilevers in liquid has been demonstrated. The contacts to the piezoresistor need to be passivated so that they do not short in conductive or ionic solutions. Using piezoresistive cantilevers in liquid is appealing because of the lack of laser alignment through the liquid. Piezoresistive cantilevers are currently designed with a high spring constant in order to increase their sensitivity. For use in contact mode on soft sample cantilevers with

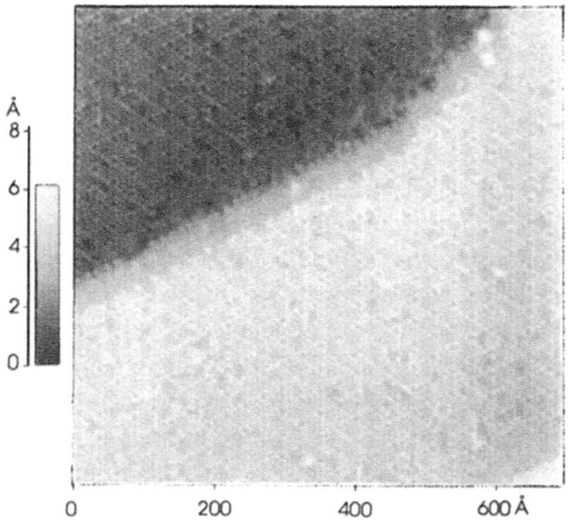

Figure 6. Image of Si(111) 7x7 collected with a piezoresistive cantilever in tunneling mode.

lower spring constant can be designed. Alternatively, modes of operation where the cantilever is vibrated should be used. AC modes of operation in liquid have been recently demonstrated.

Finally, piezoresistive cantilevers have been used for imaging of lateral forces. Minne et al.[21] have fabricated a piezoresistive cantilever with two piezoresistors located on the cantilever in such a way that both the deflection and the torsion of the cantilever can be measured simultaneously.

CONCLUSIONS

Piezoresistive cantilevers offer an alternative cantilever deflection detection scheme for AFM. This technique uses a piezoresistor integrated into the silicon cantilever. Construction and operation of the microscope is simplified because no sensor-to-cantilever alignment is required. Piezoresistive cantilevers have been used both in contact and non-contact mode, in air, liquid, UHV, low temperature, magnetic force microscopy and lateral force microscopy.

REFERENCES

1. G. Binnig, C.F. Quate, and Ch. Gerber, Atomic force microscope, *Phys. Rev. Lett.*, 56:930-933 (1986).
2. R. Erlandsson, G.M. McClelland, C.M. Mate, S. Chiang, Atomic force microscopy using optical interferometry, *J. Vac. Sci. Technol.* A, 6:266-270 (1988).
3. D. Rugar, H.J. Mamin, R. Erlandsson, J.E. Stern, B.D.Terris, Force microscope using a fiber-optic displacement sensor, *Rev. Sci. Instrum.*, 59:2337-2340 (1988).
4. T. Göddenhenrich, H. Lemke, U. Hartmann, C. Heiden, Force microscope with capacitive displacement sensor, *J. Vac. Sci. Technol.* A, 8:383-387 (1990).
5. G. Neubauer, S.R. Cohen, G.M. McClelland, C.M. Mate, Force microscopy with a bidirectional capacitance sensor, *Rev. Sci. Instrum.*, 61:2296-2308 (1990).
6. G. Meyer, N.M. Amer, Novel optical approach to atomic force microscopy, *Appl. Phys. Lett.*, 53:1045-1047 (1988).
7. R.C. Barrett, Ph.D. Dissertation, Stanford University (1991).
8. D. Sarid, P. Pax, L.Yi, S. Howells, M. Gallagher, T. Chen, V. Elings, D. Bocek,Improved atomic force microscope using a laser diode interferometer, *Rev. Sci. Instrum.*, 63:3905-3908 (1992).
9. J.Brugger, R.A. Buser, N.F. de Rooij, Micromachined atomic force microprobe with integrated capacitive read-out, *J. Micromech. Microeng.*, 2:218-220 (1992).
10. J. Tansock, C.C. Williams, Force measurement with a piezoelectric cantilever in a scanning force microscope, *Ultramicroscopy*, 42-44:1464-1469 (1992).
11. T. Itoh, T. Suga, Force sensing microcantilever using sputtered zinc oxide thin film, *Appl. Phys. Lett.*, 64:37-39 (1994).
12. C.S. Smith, Piezoresistance effect in germanium and silicon, *Phys. Rev.*, 94:42-49 (1954).
13. S. Middelhoek, Silicon sensors, Academic Press (1989).
14. M. Tortonese, R.C. Barrett, C.F. Quate, Atomic resolution with an atomic force microscope using piezoresistive detection, *Appl. Phys. Lett.*, 62:834-836 (1993).
15. F.J. Giessibl, B.M. Trafas, Piezoresistive cantilevers utilized for scanning tunneling and scanning force microscope in ultrahigh vacuum, *Rev. Sci. Instrum.*, 65:1923-1929 (1994).
16. F.J. Giessibl, Atomic force microscopy in ultrahigh vacuum, *Jpn. J. Appl. Phys.*, 33:3726-3734 (1994).
17. Y.Martin, C.C. Williams, H.K. Wickramasinghe, Atomic force microscope-force mapping and profiling on a sub 100-Å scale, *J. Appl. Phys.* 61:4723-4729 (1987).
18. Q. Zhong, D. Inniss, K. Kjoller, V.B.Elings, Fractured polymer/silica fiber surface studied by tapping mode atomic force microscopy, *Surf. Sci. Lett.*, 290:L688-L692 (1993).
19. F.J. Morin, T.H. Geballe, C. Herring, Temperature dependence of the piezoresistance of high-purity silicon and germanium, Phys. Rev. 105:525-539 (1957).
20. C.W. Yuan, E. Batalla, M. Zacher, A.L. de Lozanne, M.D. Kirk, M. Tortonese, Low temperature magnetic force microscope utilizing a piezoresistive cantilever, *Appl. Phys. Lett.*, in press (1994).
21. S. Minne, private communication.

NANOMETER-SCALE QUALITATIVE ANALYSIS OF SURFACES WITH A MODIFIED SCANNING TUNNELING MICROSCOPE/ FIELD EMISSION SOURCE

P. G. Van Patten, J. D. Noll, and M. L. Myrick

Department of Chemistry and Biochemistry
University of South Carolina
Columbia, SC 29208

Abstract: We have designed and constructed modifications to the scanning tunneling microscope that allow the bias voltage to be used as an excitation source for arc atomic emission spectroscopy. The purpose of this addition to the instrument is to allow unambiguous elemental analysis of species present on the surface. The light emitted from the arc across the tunneling gap is collected by appropriate means (such as optical fibers positioned in close proximity to the tip) and directed into a high resolution spectrometer for spectral analysis by a multichannel detector (such as a charge-coupled device (CCD) or photodiode array). Present progress includes completion of the hardware and software modifications to the instrument, verification of emission and verification that the light is substantively collected into optical fibers using a scheme described in this paper.

INTRODUCTION

The STM is used routinely to image conductor and semiconductor surfaces with atomic resolution. However, because the STM offers only topographical information, questions remain as to the identity of many structures imaged to date. This is either because of the presence of "mimic" structures (such as highly ordered pyrolytic graphite (HOPG) defects, which mimic DNA), or because the researcher has no *a priori* information on the appearance of the structure he/she seeks on the surface. Unfortunately, additional information that could provide some degree of certainty as to the identity of such structures has been lacking.

Experiments such as IR reflectance or Raman spectroscopy of the surface can give information as to the structure of an adsorbate, for example, but without a means to sample only the region directly under the tip with very high horizontal resolution, these methods cannot attest to the identity of a specific structure. Raman microscopy would perhaps be the ideal complemetary method to STM if means were available to sample the exact same spot with both techniques and if Raman spectroscopy offered low detection limits. At present, these obstacles prohibit the use of micro-Raman as an STM complement.

A good second alternative would be atomic emission spectroscopy (AES), which can overcome both of these drawbacks. Low detection limits and the adaptability of the STM experimental configuration for arc atomic emission may mean that AES is in practice the best complement to STM for identification of unknown structures under the microscope. Though structural information would still be unavailable, an elemental analysis of the surface would be useful in determining if species other than the surface material are present. One drawback–and it is a significant one–is that the technique is destructive. Thus the spatial resolution of the spectroscopic technique is limited by the smallest amount of surface which can be destroyed in acquiring a spectrum. Another drawback is that this technique is incompatible with HOPG in air due to formation of cyanogen in the arc which emits intensely throughout a broad range of the visible spectrum. This problem can be overcome with ultrahigh vacuum (UHV) STM or an inert atmosphere surrounding the head.

EXPERIMENTAL

The STM used was a Digital Instruments NanoScope II. The hardware modifications, shown in schematic form in Figure 1, included digital to analog converters (DAC's) for both the bias voltage and the Z piezoelectric crystal in parallel with their factory counterparts.

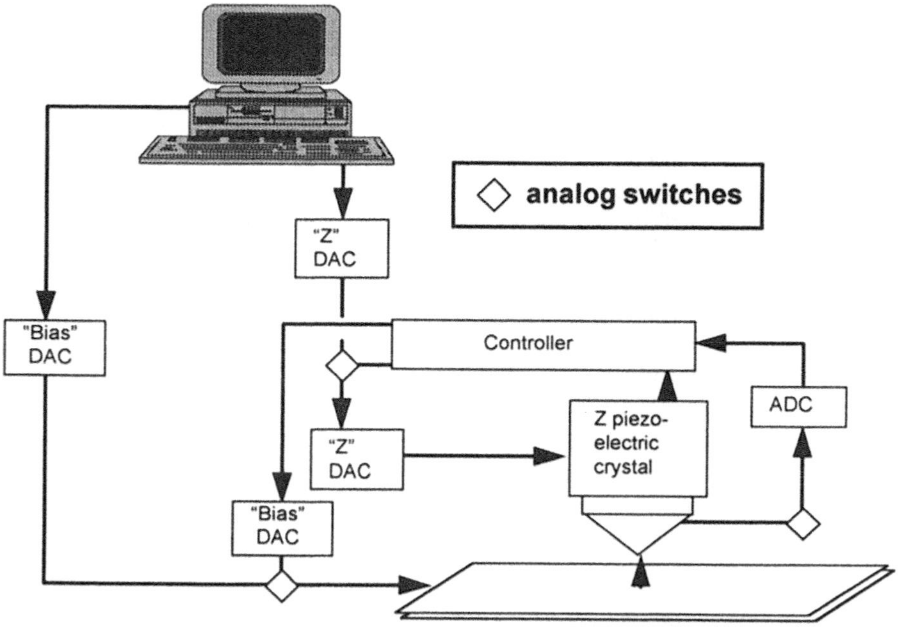

Figure 1. Schematic of the hardware modifications to the STM.

These DAC's were controlled via an IBM-compatible PC. A high-voltage amplifier was added in series with the additional bias DAC, giving a bias range of ±140 V. Analog switches were used to switch between our DAC's and the originals. Alterations were also made to the scan head to prevent damage to the tip preamplifier caused by the high voltages present. A mercury-wetted contact relay that was also controlled by the PC was inserted between the tip and tip preamp which sent the output current to the preamp (normal operation mode) or to ground (AES mode).

The program controlling the hardware modifications was written in C++. On the user's command, the program reads the voltage applied to the original Z piezo DAC, applies the same voltage to the additional Z DAC, and switches the analog switch so that the controller is unable to retract the piezo or to crash it into the surface. Next, the contact relay is switched to prevent damage to electronic components in the output path, and 2 ms is allowed for switching of the relay. Once the protective feature is in effect, the added bias DAC applies the necessary (user-selected) voltage to the input of the high voltage amplifier, and this voltage is applied to the substrate by switching the analog switch in the bias pathway. After a certain amount of time (burn time is also user-selected), the reverse process is followed to return the instrument to normal operating mode.

Optical fibers were immobilized near the tip by attaching them to an aluminum disk specially made for this purpose. The disk is designed to attach to the STM stage and to sit on top of the sample with the tip penetrating through a hole in the center. A drawing of the disk is shown in Figure 2. The fibers (0.600 mm core) were attached so that the fiber face was less than 1 mm from the tunneling gap, and were aimed downward at the gap at 60° from normal.

Tungsten tips were used throughout due to the high boiling point of that material.

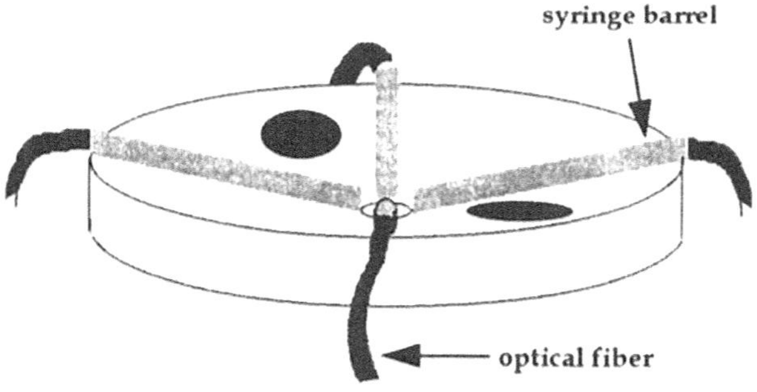

Figure 2. The mount that holds the optical fibers in position to collect the arc emission. The two dark circles are holes for the scan head supports. The small hole in the center allows the tip access to the sample surface.

RESULTS

Our preliminary results were obtained without a spectrograph by simply placing a CCD camera near the head and focusing it onto the tip. Figure 3 shows pictures obtained by the camera. In Figure 3(a), the tip holder and tip are shown above the surface. In Figure 3(b), the same area is shown without illumination in normal tunneling mode, and in the third, an intense arc is observed between the tip and surface as >100 V are applied to the surface. This arc is visible to the naked eye in the dark under low level magnification.

It is not yet known over what tip-surface separation, this arc can be sustained; however, it is known that with the tip withdrawn from the surface with the stepper motor (~ 4-5μm), the

Figure 3. (a) The STM tip holder and tip under normal room light. (b) The same picture in the dark under normal tunneling conditions. (c)Intense photon emission from the arc is evident at the tip.

arc does not initiate. With large scan heads (capable of Z displacements of over 4 μm) modified as was this one, this matter could be explored in some detail.

During the course of these experiments, the analog switch controlling the Z DAC became damaged due to a corrosion-induced current leakage, disabling our control over the Z piezo. Subsequently it was noticed that the arcs are very short-lived (perhaps 0.1s), even when the applied voltage pulse is substantially larger than this. This leads to the conclusion that either the retraction of the tip is sufficient to extinguish the arc, or that the damage incurred by the tip and surface changes the tip-surface separation enough to extinguish the arc, or that we are observing a transient spark which is unable to sustain itself for some other reason.

The next task was to collect the emitted light into optical fibers placed close to the tip. Figure 4 shows the verification that light was successfully collected by the fibers. Shown is a single 0.600 mm fiber face backlit by room light (Figure 4(a)), under normal tunneling conditions in a darkened room (Figure 4(b)), and during the high voltage arc (Figure 4(c)). The position of the fiber was identical in all three pictures, and the same as that described above.

Figure 4. (a) 0.600 mm fiber face backlit by white light. (b) The same fiber face in the dark. (c) The fiber backlit by the arc emission.

The present priority is acquisition of an atomic spectrum from the copper substrate. Further research will be aimed at optimizing the arc parameters for resolution, intensity, and background suppression. The pertinent parameters will include (but not necessarily be limited to) the maximum applied voltage, the voltage as a function of time (and thus the burn time), the tip-surface separation, and the tip material.

ATOMIC FORCE MICROSCOPY IMAGING OF SINGLE ION IMPACTS ON MICA

D. C. Parks,[1*] R. Bastasz,[3] R. W. Schmieder,[2] and M. Stöckli[1]

[1]Macdonald Laboratory, Kansas State University
Manhattan, KS 66506-2601
[*]*Currently:* National Insititute of Standards andTechnology
Gaithersburg, MD 20899
[3]Sandia National Laboratories
Livermore, CA 94551-0969

Abstract: The electron affinity and potential energy that is associated with highly charged ions may be used to create nanometer scale damage sites on the surface of insulating materials. We have used atomic force microscopy to image the surface damage caused by single ion impacts. Freshly cleaved mica was irradiated by low energy Xe^{44+} ions at normal incidence. Impact sites are typically circular protrusions 20 nm in diameter and 0.3 nm in height. Lateral force microscopy shows the damage sites to have increased friction relative to the surrounding undisturbed crystal.

INTRODUCTION

The interaction of ions with surfaces is widely used in applications such as sputter cleaning and ion implantation. These can lead to a variety of technologically useful physical and/or chemical modifications of surface and near surface regions.

Swift ions carry large amounts of energy, but principally interact by way of electronic inelastic scattering with matter, at least until near the limit of their penetration range. Their energy is distributed along the ion track. The deposited energy density then approaches an upper bound which in the high velocity limit is only weakly dependent on ion energy. We use slowly moving, but highly charged ions to deposit a large amount of energy near the surface of insulating materials. These ions can modify the surface by ion neutralization processes and deposition of the associated coulomb energy. The deposited energy density may then be larger then that available from swiftly moving ions.

Insulators may be most susceptible to local lattice damage by these means due to their low electronic mobility. This is especially important when the damage is mainly due to local

Atomic Force Microscopy/Scanning Tunneling Microscopy 2
Edited by S.H. Cohen and M.L. Lightbody, Plenum Press, New York, 1997

charge depletion. Damage may occur when the ion is quickly neutralized by lattice electrons and the crystal subsequently repels itself in a sort of "coulomb explosion."[1]

Scanned probe microscopy has been used to image ion-damaged surfaces for several ion-surface combinations. Surface roughening of Si[2] and highly ordered pyrolytic graphite (HOPG)[3] surfaces as measured by STM and individual ion impacts on semiconductors[4,5] and HOPG[6-10] have been observed using a variety of singly charged ions. Etched[11] and unetched[12] ion impact sites on insulating surfaces are observable by AFM.

EXPERIMENT

Freshly cleaved muscovite mica was chosen as the target material because of its high resistivity and its cleavability to produce large atomically flat regions. 8 mm × 8 mm samples mounted on 12 mm stainless steel discs were irradiated at normal incidence under UHV with the target chamber base pressure ~3 × 10^{-9} torr. The mica surface was masked by a 3 mm diameter aperture. A thin spacer ring protected the mica surface from contact scratches. This arrangement provided a pristine unirradiated region on each sample for control measurements.

The KSU Cryogenic Electron Beam Ion Source (Cry-EBIS) was used to produce a $^{136}Xe^{44+}$ ion beam. These ions carry ~51 keV potential energy. The 3.5 keV/q kinetic energy as supplied by the source could be varied from 0.10 keV/q to 10 keV/q at the target by a deceleration/acceleration lens. The time averaged ion current was approximately 5 pA delivered in 10 ms long pulses at ~5 Hz, giving a flux of ~0.1 ions/s·μm^2. Samples were irradiated to give a total fluence of approximately 10^9-10^{10} ions/cm^2. This density gives 10-100 ion impact sites/μm,[2] which is a high enough density to unambiguously identify individual ion impact sites within a 1 μm × 1 μm AFM scan region, but not so high as to have a significant number of overlaps.

Samples were imaged by constant deflection contact mode AFM in ambient conditions using a NanoScope II[13] with Si_3N_4 tips. The scanning force was adjusted so as to be near minimum, but yet stable. Typical scans were 1 μm × 1 μm, acquired at 5 scan lines/s. All samples were scanned in four directions: 0°, 90°, 180°, 270°. This was done to identify possible direction induced scanning artifacts.

Control scans of regions behind the mask of each sample were also obtained. None of the control area scans show the characteristic single ion impact features. Typical unirradiated mica is flat and featureless, with noise ~0.1 nm, when imaged at 1 μm × 1 μm scan size.

RESULTS

AFM images of the irradiated mica reveal individual ion impact sites. These sites are imaged as raised circular cones. The measured density of impact sites and the expected number as calculated from the ion fluence are in reasonable agreement. Accurate measurement of the ion current and nonuniformities of the beam profile make precise damage efficiency calculations impossible.

Figure 1 shows a 1 μm × 1 μm contact mode AFM scan of an irradiated area on mica. Ion impact sites are visible as the small white dots. Typical sizes of these sites are measured to be 20 nm in diameter and 0.3 nm high. This is an example of the typical image obtained in this experiment. However, some scans reveal the ion impact sites to be indentions rather than protrusions. This is evident in the 100 nm × 100 nm contact mode AFM scan shown in Figure 2. Here are four ion impact sites of approximately the same dimensions as noted above but with reversed topography.

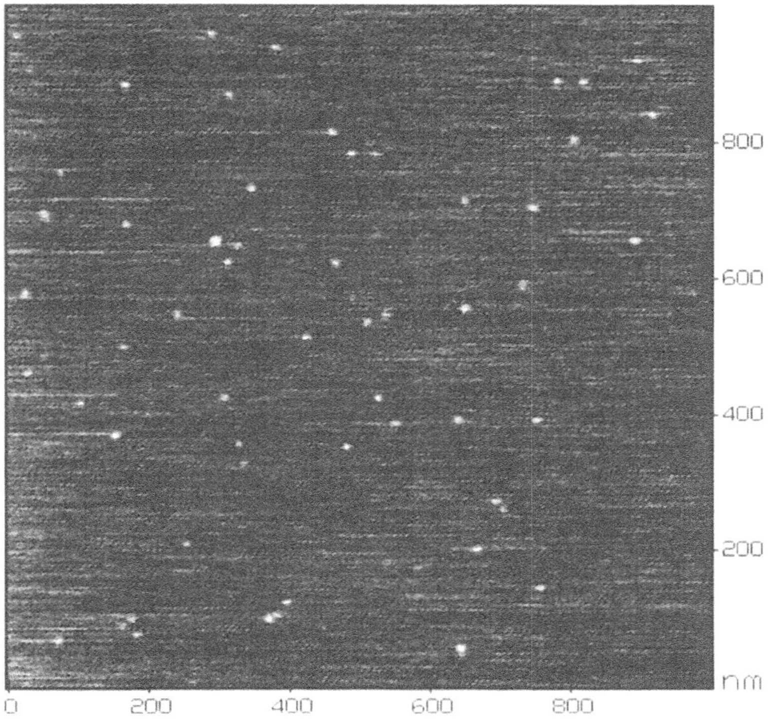

Figure 1. Contact mode atomic force microscope image of mica irradiated by Xe^{44+} ions. Individual ion impact sites are seen as white dots. Typical feature size is 20 nm diameter and 0.3 nm height. Image size is 1 μm × 1 μm.

The reversal of topography was only noted for some scans made in the 0° and 90° directions. Adjustment of applied force on the tip did not produce any effect on measured height or cause systematic reversal of topography. A few samples were imaged using a NanoScope III[13]. This allowed us to obtain tapping mode and lateral force mode information. We could also simultaneously obtain trace and retrace information for a given contact mode image. When data from the leftward and rightward moving segments of the scan were retained, the protrusions were shown to reverse apparent topography. This indicates that part of the topography signal is due to laterally varying friction between the probe and the surface.

Lateral force microscopy shows the increased torsion on the tip when travelling over an ion impact site (Figure 3). The lateral friction is proportional to the sum of the applied load on the tip and attractive force between the tip and surface. The varying attractive interaction between the tip and surface as well as differing coefficient of friction both contribute to the observed signal. The ion impact sites correspond to locations of increased lateral friction and/or increased tip adhesion to the surface. This increase in friction has also been noted for swift ion impact sites.[12,14]

Cross sections through individual ion impact sites (Figure 4) reveal a roughly triangular profile. Diameters are roughly 10-20 nm with 0.3-0.4 nm heights. The two narrowly separated pits demonstrate that features spaced as closely as 12 nm can be resolved. Sizes of pit type features may be slightly underestimated due to finite tip radius on the probe. Similarly, sizes of protrusions may be slightly over estimated. However, because of the very low aspect ratio, the finite tip radius should have minimal effect on the observed features.

To ascertain if the amount of observable crystal damage is a function of ion kinetic energy, the ion kinetic energy was varied in the range 0.1 - 10 keV/q. Alternatively, varying the interaction time between the ion and the near surface regions just prior to and immediately after impact could also change the amount of damage. Both diameter and height of the surface damage were measured for several different energies (Figure 5). The observed surface features are not seen to change within the range of energies studied. On some samples pits were seen (plotted as negative heights). There is no correlation between observation of pits and ion

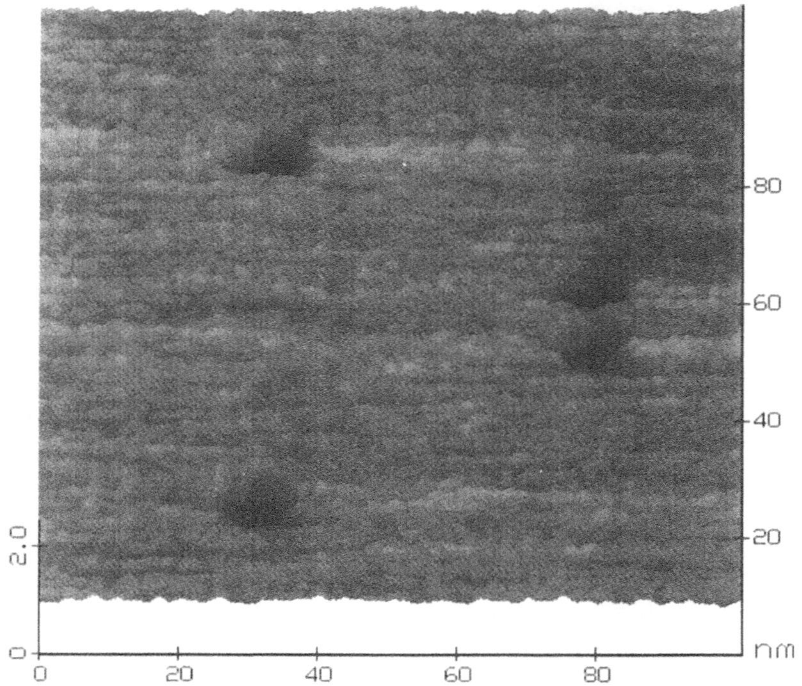

Figure 2. Pit type single ion impact sites. This topographic inversion is sometimes seen both in contact mode and in tapping mode Atomic Force Microscopy. In contact mode it is only seen in scans traveling in the 0° and 90° directions. Image size is 100 nm × 100 nm.

kinetic energy. However, pits were only seen for scans travelling in the 0° and 90° directions. This suggests that it may be a friction related phenomenon. Pits were seen with several different tips and on several different samples.

These ion damage sites are easily disturbed by observation with contact mode AFM. The signal of single ion impacts was observed to gradually diminish in contrast upon scanning. With scan areas of 1 μm × 1 μm, approximately 10 consecutive scans in a given area was sufficient to render all signal to a level approximately equal to the noise. This process was much quicker

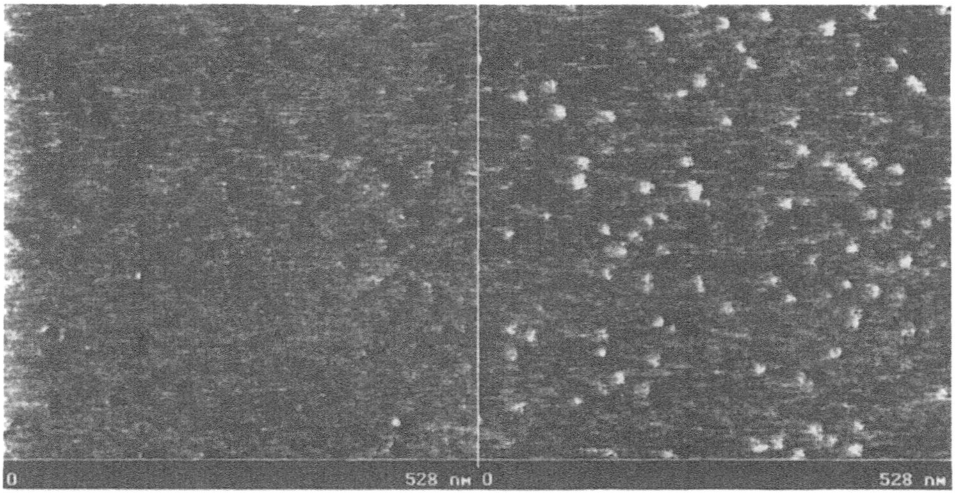

Figure 3. Lateral force microscopy image of mica irradiated with Xe^{44+} ions. The two images were obtained simultaneously, keeping information from both the left-traveling and the right-traveling parts of the scan. The signal reversal upon reversal of scan direction shows varying lateral forces. The ion impact sites are seen as regions of higher tip-surface attraction forces and/or increased coefficient of friction.

at smaller scan sizes, but not very sensitive to tip velocity. The obvious increased scan line density must be an important factor in the rate of signal deterioration. Scans of reduced size but also reduced scan line densities show an extended durability under otherwise equal scanning. The effect can be clearly seen in Figure 6. The center of this 3 μm × 3 μm AFM image was scanned at 1 μm × 1 μm rotated 45°. Then another 1 μm × 1 μm region at center bottom was scanned. The flattening of the closely spaced ion impact sites in the previously scanned regions is clear. The absence of ejecta at the edges of these regions suggests that the effect is not one of removal of material.

DISCUSSION

The interaction of highly charged ions with surfaces begins when the electric field of the ion begins to polarize the material surface. The beginning of local charge depletion in insulating surfaces occurs even before ion impact. When the ion-surface separation is sufficiently small, (~30 nm) electrons may be removed from the crystal by field emission processes. These electrons initially are captured into high lying Rydberg states, forming a "hollow atom." The ion continues to capture more electrons as it moves closer, and ejects auger electrons as the high lying states begin their cascade to the ground state. Since the time from first electron capture to ion impact is not much longer than the auger decay time, there is not sufficient time to completely neutralize the ion or to complete the auger cascades.

When the ion impacts on the surface, electrons are now available to directly fill inner shell vacancies from the crystal lattice. This neutralization completes the local charge depletion. The now neutralized ion keeps moving through the lattice until slowed by inelastic electron interactions and eventually stopped within the bulk crystal. The charge depleted region of the

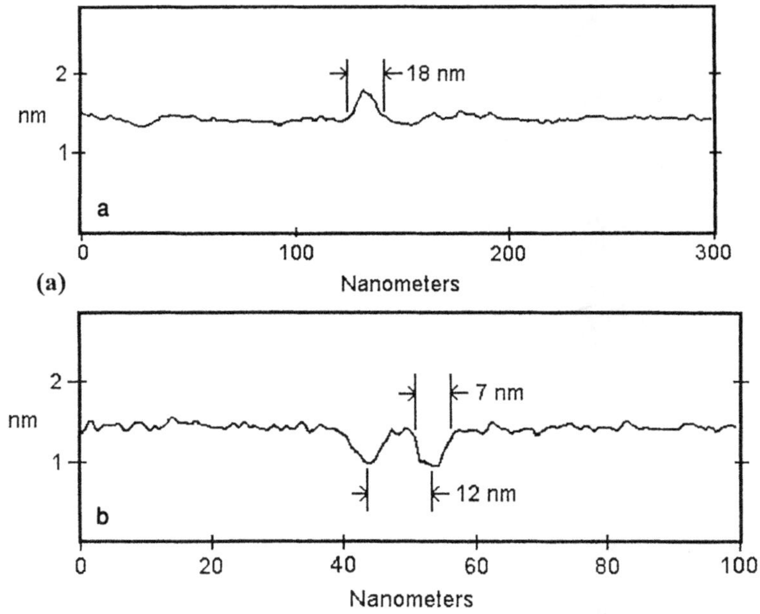

Figure 4. Cross sections through contact mode AFM images of individual ion impact sites. A blister type feature is shown in (a) with a 18 nm diameter and 0.4 nm height. The ~0.1 nm background noise is shown. Two closely spaced pit type features (b) indicate the lateral resolution (~12 nm) possible between adjacent features.

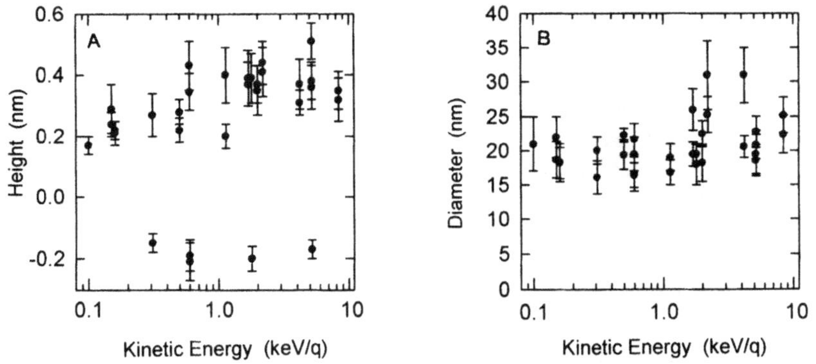

Figure 5. Height (a) and diameter (b) of the same set of individual ion impact sites as measured by contact mode AFM as a function of ion kinetic energy. Pits are plotted as negative heights in (a) but undifferentiated in (b). There is no systematic trend in either size measurement when varying the ion kinetic energy.

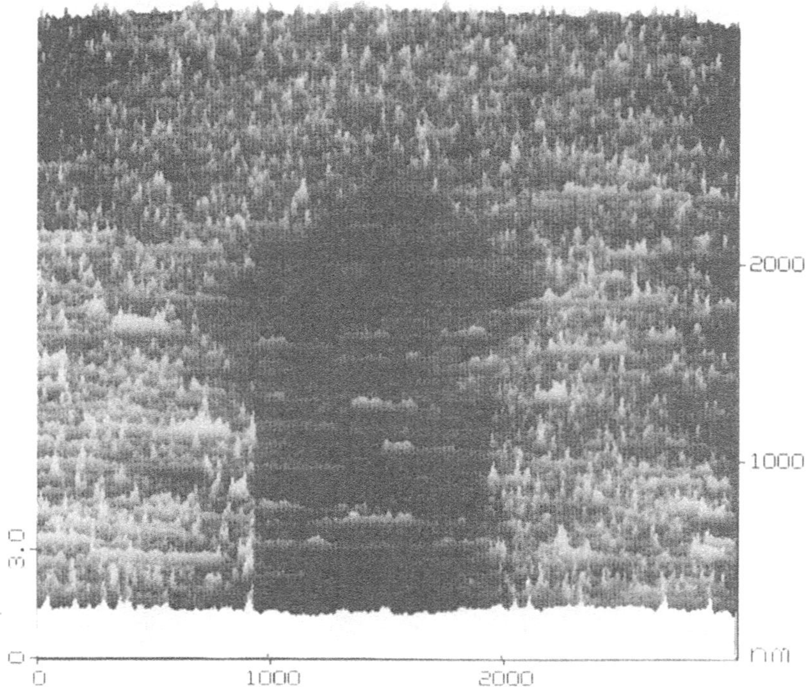

Figure 6. Continuous previous scanning at a 1 μm × 1 μm scan size at the center (rotated 45°) and center bottom (rotated 0°) flattened the high density ion impact features. The absence of debris at the edges of the smaller scan regions indicates a flattening rather than a scraping action. Image size is 3 μm × 3 μm.

crystal lattice is now locally self-repulsive. The response of the lattice is to break the relatively weak bonds between the mica layers. This anisotropic lattice expansion is observed as local rises at the surface.

Upon scanning, these protrusions are relatively easily flattened. With small radius probes, the local pressure is sufficient to partially collapse these relatively soft regions. Increased scan line density increases the damage density, which hastens the flattening process. Single ion damage sites are observed to be stable and observable when stored in air for several months.

The ambiguity of the true topographic nature of the ion impact sites still exists. The interaction of the probes with the surface clearly is an important factor. More investigations with different tip materials, lateral force microscopy and tapping or noncontact imaging modes are needed to help resolve this uncertainty.

ACKNOWLEDGMENTS

This work was supported by the U. S. Department of Energy under contract DE-AC04-94Al85000 and by the Division of Chemical Sciences, Office of Basic Energy Sciences, Office of Energy Research, U.S. Department of Energy.

REFERENCES

1. E. S. Parilis, L. M. Kishinevsky, N. Yu. Turaev, B. E. Baklitzky, F. F. Umarov, V. K. Verleger, S. L. Nizhnaya, and I. S. Bitensky, Atomic Collisions on Solid Surfaces, North Holland, Amsterdam, 539 (1993).
2. R. M. Feenstra and G. S. Oehrlein, Surface morphology of oxidized and ion-etched silicon by scanning tunneling microscopy, *Appl. Phys. Lett.* 47:97-99 (1985).
3. E. A. Eklund, R. Bruinsma, J. Rudnick, and R. S. Williams, Submicron-scale surface roughening induced by ion bombardment, *Phys. Rev. Lett.* 67:1759-1762 (1991).
4. H. Feil, H. J. W. Anadfliet, M.-H. Tsai, J. D. Dow and I. S. T. Tsong, Random and ordered defects on ion-bombarded Si(100)-(2×1) surfaces, *Phys. Rev. Lett.* 69:3076-3079 (1992).
5. I. H. Wilson, N. J. Zheng, U. Knipping, and I. S. T. Tsong, Scanning tunneling microscopy of ion impacts on semiconductor surfaces, *J. Vac. Sci. Technol. A* 7:2840-2844 (1989).
6. L. Porte, C. H. de Villeneuve and M. Phaner, Scanning tunneling microscopy observation of local damages induced on graphite surfaces by ion implantation, *J. Vac. Sci. Technol. B* 9:1064-1067 (1990).
7. R. Coratger, A. Claverie, F. Ajustron, and J. Beauvillian, Scanning tunneling microscopy of defects induced by carbon bombardment on graphite surfaces, *Surface Science* 227:7-14 (1990).
8. G. M. Shedd and P. E. Russell, The effects of low-energy ion impacts on graphite observed by scanning tunneling microscopy, *J. Vac. Sci. Technol A* 9:1261-1264 (1991).
9. R. Coratger, A. Claverie, A. Chahboun, V. Landry, F. Ajustron, and J. Beauvillian, Effects of ion mass and energy on the damage induced by an ion beam on graphite surfaces: a scanning tunneling microscopy study, *SurfaceScience* 262:208-218 (1992).
10. S. Bouffard, J. Cousty, Y. Pennec, and F. Thilbaudau, STM and AFM observations of latent tracks, *Radiat. Eff. and Defects in Solids* 126:225-228 (1993).
11. D. Snowden-Ifft, P. B. Price, L. A. Nagahara, and A. Fujishima, Atomic-force-microscopic observations of dissolution of mica at sites penetrated by keV/nucleon ions, *Phys. Rev. Lett.* 70:2348-2351 (1993).
12. F. Thibaudau, J. Cousty, E. Balanzat, and S. Bouffard, Atomic-force-microscopy observations of tracks induced by swift Kr ions in mica, *Phys. Rev. Lett.* 67:1582-1585 (1991).
13. Digital Instruments, Santa Barbara, CA
14. T. Hagen, J. Ackerman, N. Angert, S. Grafstrom, M. Neitzert, R. Neumann, C. Trautmann and J. Vetter, Friction studies of heavy-ion irradiated mica on a sub-μm scale using a scanning force microscope, *GSI Scientific Report 1992,* 303 (1992).

AFM/STM IN MATERIALS SCIENCE

Moderators: Ernest C. Hammond, Morgan State University
Timothy L. Porter, Northern Arizona University

APPLICATIONS OF ATOMIC FORCE MICROSCOPY IN OPTICAL FIBER RESEARCH

Q. Zhong[1] and D. Inniss[2]

[1]AT&T
Holmdel, NJ 07733
[2]Lucent Technologies
Murray Hill, NJ 07974

Abstract: In this paper, novel applications of atomic force microscopy (AFM) to optical fiber research are reviewed. Three specific examples are presented, illustrating the effective use of AFM to advance our understanding of material structure and properties in optical fibers. Existing issues and the need for innovation in applications of AFM to new material systems are also discussed.

INTRODUCTION

Optical fibers are the building blocks of telecommunication systems upon which the National Information Infrastructure relies. The reliability and performance of optical fibers are, therefore, critical issues of technological importance. From a material point of view, an optical fiber comprises silica glass fiber, which typically is about 125 μm in diameter, consisting of the core and the clad region and of the polymer coating, which includes a low-modulus UV-curable urethane acrylate as the primary coating and a high-modulus acrylate as the secondary coating for a dual-coat fiber. Figure 1 illustrates a typical design for lightguide fibers.

Performance issues of concern in optical fiber applications include mechanical reliability of fibers, glass/polymer interfacial adhesion, and optical properties such as loss associated with transmission or fiber interconnect. Various aspects of the glass/polymer composite are responsible for these properties, and they are, to a certain extent, interrelated. For example, glass fibers placed in hot and humid environment, which is not an unusual service condition, may result in strength reduction to an unacceptable level. This phenomenon, known as aging or corrosion of glass, has been the subject of many studies.[1] On the other hand, it was found that poor adhesion at glass/polymer interfaces may accelerate the aging of glass fibers, resulting in accelerated strength degradation.[2]

Atomic Force Microscopy/Scanning Tunneling Microscopy 2
Edited by S.H. Cohen and M.L. Lightbody, Plenum Press, New York, 1997

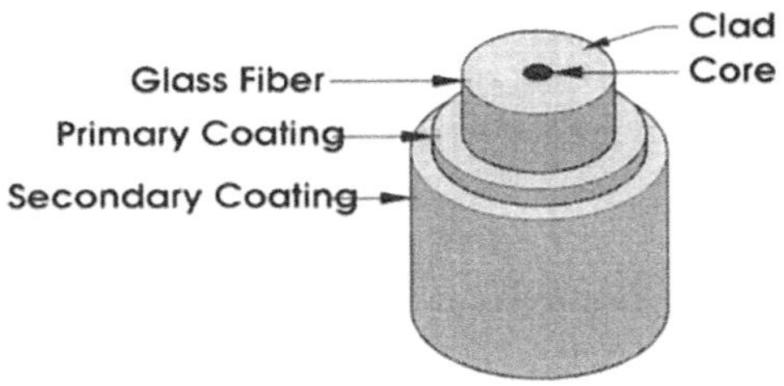

Figure 1. Schematic design for a typical optical lightguide fiber.

The success of delineating these material properties depends upon the ability to characterize the glass and polymer, especially surfaces and interfaces, to which these properties are intimately related. Since the glass and polymers are usually dielectric materials, direct use of many surface characterization techniques may be hampered. Atomic force microscopy (AFM), on the other hand, emerges as a technique well suited for these purposes. AFM allows direct study of dielectric materials without the need of a conductive coating. In addition, AFM provides a powerful local probe to study, on a nanometric scale, mechanical interactions at surfaces. The application of AFM to optical fiber research was initiated in the scope of examining glass surface flaws.[3,4] This novel approach was well appreciated in the optical fiber research community as it provided an experimental input to the mechanical reliability assessment of fiber lifetime. Although the initial model presents inconsistency with the experimental data, our knowledge base in the scope of glass surface flaws has advanced.

In this paper, we review an area of recent development: the application of AFM to study various aspects of the optical fiber, extending beyond the surface flaw characterization. First, adhesion of polymer/glass interfaces is addressed with the combined use of tapping mode and contact mode AFM. These studies lead to a fundamental model with which different adhesion mechanisms can be designed and eventually implemented. Second, the use of AFM to characterize surface microstructure, including surface flaws, is discussed. This is an important topic in that the aging behavior of the glass is directly responsible for the long-term mechanical reliability of optical fiber systems. In the third example, the high spatial resolution of AFM is taken advantage of to investigate the core/clad region of the fiber, where features of interests are generally sub-μm in dimension. These studies have facilitated designs and procedures to improve the performance of optical fibers.

These examples serve to illustrate the versatility of AFM techniques. This paper intends to discuss various aspects of AFM applications in optical fiber research. The cited references provide details of material aspects of the problems and solutions that this technique has provided.

CASE STUDIES

Application of AFM to Polymer/Glass Interfacial Adhesion

The novel aspect of applying AFM to study adhesion at interfaces is that it allows a nanometric characterization of the fracture surface, i.e., fracture topography and locus of failure. Polymer coating is mechanically removed from the glass fiber, and the residual polymer on the glass surface is characterized by tapping mode AFM. This yields the topography of the fracture surface from which the locus of failure at the interface can be determined. The knowledge about the interface advances with the scope of information available. As the scale of measurements approaches the molecular level, it is evident that further detailed knowledge about the interfacial fracture process can be obtained. For example, an optically clean fiber surface (i.e., an optically clean glass fiber surface is defined herein as the one that is free of polymer residues under an optical microscope with a magnification of, for example, 800x) may have surface topography as illustrated in Figure 2, revealed by AFM analysis. The surface contains polymer residues of various sizes, ranging from several hundreds of nanometers to several μm in diameter, and tens of nanometers to a few hundred nanometers in height. Clearly, the level of details revealed by an optical microscope vs. AFM is drastically different. This information is critical in determining the "true" fracture path, which is ultimately determined by adhesive vs. cohesive strength of the interface region.

The successful application of AFM to polymeric materials was initially challenged by the fact that in the conventional mode of operation (contact mode AFM), the probe tip is in continuous contact with the surface, exerting forces as high as hundreds of nN's on the polymer. This can cause distortion or even damage in the soft, viscoelastic polymer.[5] The key to polymer coating characterization is, therefore, the reproducibility of the characterization technique. With the tapping mode, the tapping force can be minimized to as low as 0.1 nN, nondestructive to the fragile polymer. We have discussed in reference 5 the successful use of tapping mode AFM to study polymeric materials, yielding reproducible surface topography (i.e., the size and distribution of the polymer residues). It is noted that although the use of tapping mode AFM is essential in characterizing the surface topography of polymer, the destructive nature of contact mode AFM can also be utilized in a positive way. Table 1 provides a summary of the kind of information available in a typical study.

Table 1. Summary of Information Obtained in an AFM Study

	Observations of Residual Polymer	Implications to Polymer/Glass Interfaces
Tapping Mode AFM	• morphology • distribution	• fracture mode • locus of failure
Contact Mode AFM	• deformation • resistance to lateral forces	• polymer deformability • bonding strength

AFM characterizations of polymer/glass interfaces can benefit our knowledge of adhesion, which ultimately leads to a better coating design strategy, and therefore, desirable performance. Two competing factors in adhesion need to be considered. On one hand, good adhesion is desirable for the coating/fiber interface to minimize potential spontaneous delamination of the coating under service conditions. On the other hand, coating removal is a necessary procedure before fiber splicing in field applications. To facilitate optimal designs, various coating/glass interfaces were developed via coating chemistry and/or physical treatment of the coating. The critical contribution of AFM therein is to assess the interface by analyzing fracture surface topography after the coating was removed in a manner similar to the field procedure. Figure 2 shows, in fact, two contrasting examples where each fracture path is distinctly different from the other. A systematic study of various coating systems led to a generalization of the two fracture patterns, as illustrated schematically in Figure 3. Based on experimental observations, a homogeneous cohesive fracture appears to favor a clean coating removal. This example serves to demonstrate the role of a successful AFM analysis in the coating design process.

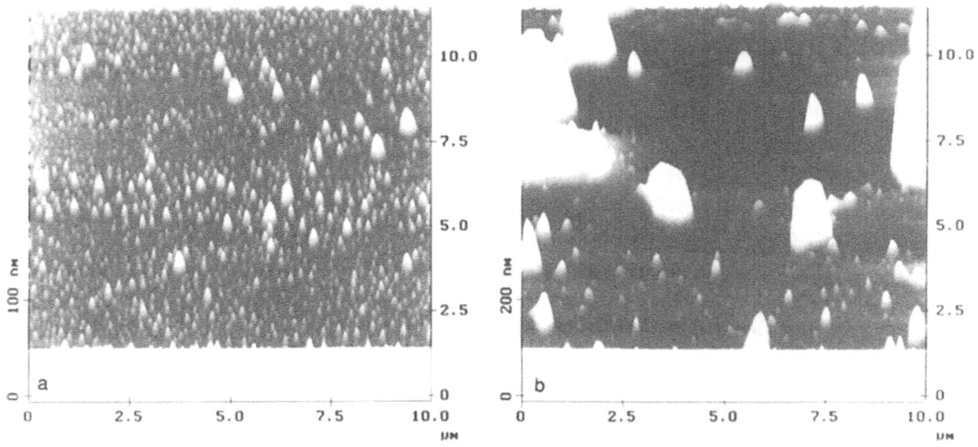

Figure 2. AFM micrographs for residual polymer on glass fiber surfaces after the mechanical removal of the coating.

Application of AFM Techniques to Studying Surface Corrosion of Glass

Microstructure of silica glass surfaces is responsible for glass strength, adhesion, corrosion resistance, and many other material properties. It has long been speculated that glass surfaces contain sharp flaws that are on the scale of nanometers.[6] Previous studies using either scanning tunneling microscopy or AFM have attributed varying degree of surface roughness to flaws responsible for the weakening of the glass.[3] However, the flaw geometry (i.e., aspect ratio or width/length) was not clearly presented. The lack of a well-defined flaw on a generally rough surface hampered such efforts. In an earlier report,[4] we have demonstrated a model system, in which flaws of well-defined geometry and distribution were developed by exposing glass fiber surfaces to the vapor of dilute hydrofluoric acid. The application of AFM to characterize the individual flaw on the fiber surface is pivotal to the proposal of a strength-flaw relationship.[4] Figure 4 shows a typical surface topography of corroded fiber on which well-defined corrosion pits are identified. The flaw geometry is measured via profiling the surface. The important conclusion of this study is that flaws are in general blunt, not sharp as previously speculated (cf. Figure 4).

An important aspect for a reproducible AFM characterization of silica surfaces is the stability of the probe tip. Observations of instability (attractive or repulsive interactions) between the probe tip and silica (including oxidized silicon) is not uncommon, especially when the silicon nitride tip is used. We have also studied the wear of silicon nitride tip in scanning silica surfaces and found that the degradation of the image resolution is obvious in a matter of one scan. By examining the tip using a scanning electron microscope, it appears that the apex of the tip was flattened significantly. Although the exact nature of the interaction is not known, electrostatic interaction appears to be responsible on numerous occasions. Attempts to minimize the interaction via charge neutralization or the use of fluids with various

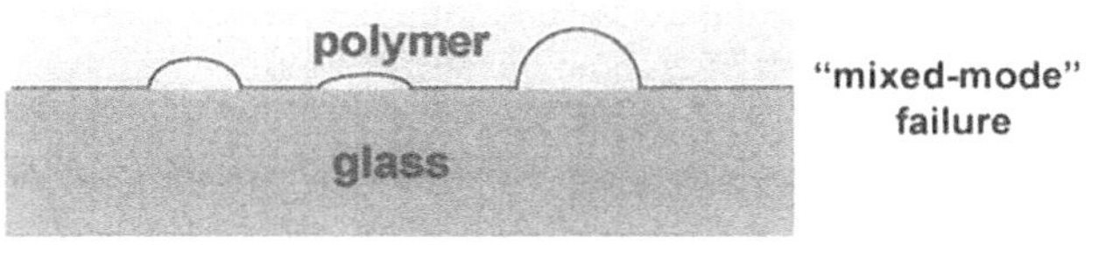

$E_f = \theta\, W_a + (1\text{-}\theta)\, W_c + W_d$

$E_f = W_c + W_d$

Figure 3. A schematic illustration of the two proposed fracture paths corresponding to the surface topography shown in Figure 2. Note that E_f denotes the fracture energy and W_a, W_c, and W_d, represents the work of adhesion, cohesion, and deformation, respectively.

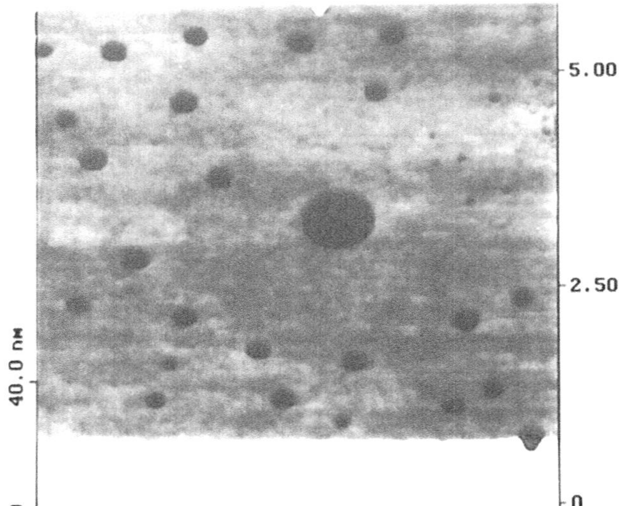

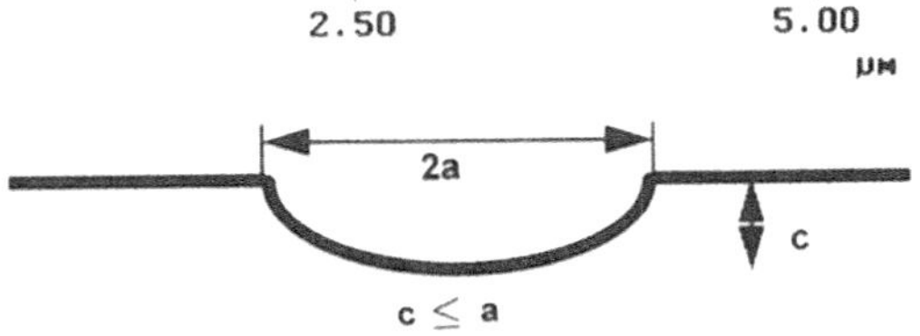

Figure 4. An AFM micrograph for a corroded glass fiber surface (exposed to the vapor of 1 vol.% HF solution), and schematic illustration of the blunt flaw.

pHs yielded varying degrees of success. However, these procedures can be difficult to implement.

A significant observation is that by using tapping mode AFM, imaging the silica surface is apparently less problematic. Resonant oscillation of the cantilever, modulated externally via a piezoelectric fixture, appears to overcome the instability of the tip when the tip is in the close vicinity of the silica surface. And the tip apex can be maintained for a reasonably long time if the oscillation amplitude (tapping amplitude) is properly adjusted. This ensures the reproducibility of the roughness measurement in the microstructure characterization.

Application of AFM to Study Core/Clad Structure

Increasing applications of specialty fibers, which typically are small core fibers, often present a need to characterize the structure and geometry of the core region with high resolution. As the utilization of AFM in optical fiber research evolves, it seems prudent to extend its virtue of being a local probe to study the core region of the fiber. We have developed a procedure[7] to delineate the structure at the core/clad region. The fiber of interest is first cleaved followed by etching in a solution, such as dilute hydrofluoric acid. The purpose of the etch is two-fold: to expose the core structure resulting from differential etching (due, for example, to different compositions), and to eliminate the electrostatic charge that is apparently associated with the as-cleaved surface. AFM is then used as a local probe to profile the fiber endface. As an example, Figure 5 illustrates several different fibers profiled.

The novelty of using AFM in studying core region is that AFM provides superior high 3-D resolution. A number of other analytical techniques, most notably scanning electron microscopy, have superior lateral resolutions but no direct measure of depth because of their reliance on the contrast mechanisms. AFM, on the other hand, can provide a direct measure of height variations with resolution attainable on the order of subnanometers. This feature of high resolution readily provides measurements such as core diameter or core ovality. In addition to the microstructural and geometrical information about the core, one of the significant applications is that the refractive index variation in the core/clad region can be inferred from the etching depth profile. This application is based upon the fact that the index change is related to the germanium doping content, which, in turn, affects the etching rate. The different etching rate, in effect, provides a spatial correlation between the refractive index and the microstructure and/or geometry of the fiber endface. The area becomes attractive for continuing research where refractive index differences in a small core fiber can be resolved, and the segregation of different dopants may be located in the core region, provided a differential etching rate can be obtained.

SUMMARY

It is imperative to note that, despite increasing success in applying AFM to various aspects of optical fiber research, there are limitations pertaining to the technique. Most notably the AFM does not provide chemical information, except in limited cases where chemical identities can be inferred from indirect measures. It should also be noted that no resolution of a single atom is yet possible with AFM, which largely can be attributed to the lack of a single-atom probe tip. This is not to be confused with the resolution of periodic

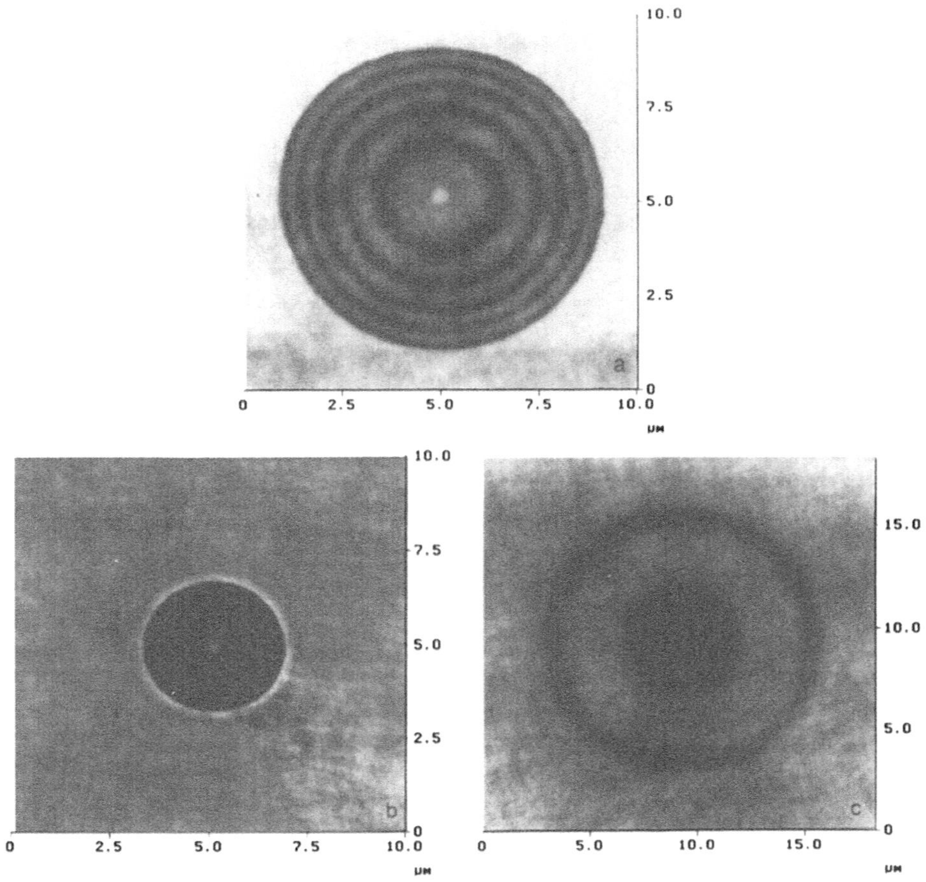

Figure 5. AFM micrographs for the endface of various fibers after the fibers were cleaved and etched in dilute HF solution.

structure, such as the resolution of the atomic structure of mica. Specifically, in applications of polymer or glass, our ability to resolve subnanometric structure is still rather limited because of the absence of ordered structure and the resolution limited by the tip radius. Other factors including the common practice of ambient imaging cannot preclude the possibility of surface contamination. These remarks are certainly more general than their applicability to optical fiber research, so is the need to search for innovative ways of utilizing the technique in various new material systems.

ACKNOWLEDGMENT

The authors would like to acknowledge many of their colleagues for their invaluable input to this study.

REFERENCES

1. C.R. Kurkjian and D. Inniss, Understanding mechanical properties of lightguides: a commentary, *Opt. Eng.*, 30 (6), 681-89 (1991).
2. H.C. Chandan, J. R. Petisce, J. W. Shea, C. R. Taylor, L. L. Blyler, D. Inniss, and L. Shepherd, Fiber protective coating design for evolving telecommunication applications, *Int'l. Wire & Cable Sym. Proc.*, 239-48 (1992).
3. R. S. Robinson and H. H. Yuce, Scanning tunneling microscopy of optical fiber corrosion: surface roughness contribution to zero-stress aging, *J. Am. Ceram. Soc.*, 74(4), 814-18 (1991).
4. D. Inniss, Q. Zhong, and C. R. Kurkjian, Chemically corroded pristine silica fibers: blunt or sharp Flaws?, *J. Am. Ceram. Soc.*, 76(12), 3173-3177 (1993).
5. Q. Zhong, D. Inniss, K. Kjoller, and V. Elings, Fractured polymer/silica interface studied by tapping mode atomic force microscopy, *Surf. Sci.*, 290, L688-92 (1993).
6. A. A. Griffith, The phenomena of rupture and flow in solids, *Philos. Trans. R. Soc. London*, A, 221, 163-98 (1920).
7. Q. Zhong and D. Inniss, Atomic force microscopy: characterizations of the lightguiding structure of optical fibers, *J. Lightwave Technol.*, 12(9), 1517-1523 (1994).

ATOMIC FORCE MICROSCOPY STUDIES ON OPTICAL FIBERS

M. John Matthewson,[1*] Vincenzo V. Rondinella,[1,2] and James Colaizzi[1]

[1]Fiber Optic Materials Research Program
Department of Ceramic Science and Engineering
Rutgers University P. O. Box 909
Piscataway, New Jersey 08855
[2]Present address: Institute for Transuranic Elements
Postfach 2340, 76125 - Karlsruhe
Germany
**Correspondence addressee*

Abstract: The strength of brittle materials is controlled by the presence of stress concentrating defects which can be produced during manufacture or subsequent use. Surface defects, which dominate for glasses, can slowly grow under the combined influence of applied stress and environmental moisture, leading to delayed failure. However, the pristine as-drawn surface of optical fibers is preserved by the application of a polymer coating immediately after drawing so that defects are of atomic dimension. For such high strength material (1GPa), the strength can be degraded in aggressive environments by surface corrosion. AFM has proved a valuable tool for examining changes in the surface morphology of optical fibers on the nanometer scale and relating them to strength degradation. The surface of fused silica fiber corrodes in aggressive (*i.e.*, warm and wet) environments to form roughness which, while only a few nm root mean square (rms), can substantially degrade the strength leading to drastically foreshortened lifetimes. The incorporation of nanosized silica particles in the fiber coating extends the lifetime under stress by factors of up to 300 or more. AFM has verified that the particles operate by slowing the surface roughening process. Heavy metal fluoride glass fibers (based on "ZBLAN" compositions) have many potential applications but are considerably less durable than fused silica and can rapidly degrade in room temperature water. Surface analysis using AFM has shown that the fastest degradation is brought about by conditions that lead to the formation of crystals in the fiber surface.

INTRODUCTION

The strength of brittle materials is controlled by the presence of submicroscopic defects that concentrate the applied stress. These defects are normally in the surface of oxide glasses and result from contamination and handling damage introduced during production and use. While not directly observable, the presence of these defects may be inferred since the strength of such materials is broadly scattered and is orders of magnitude lower than the theoretical strength calculated from chemical bond strengths.

Fused silica glass fibers are widely used for optical communications. The fibers are 125 μm in diameter and light is guided down their central core (typically 6 μm diameter for single mode fiber) which is chemically doped to produce slightly different optical properties. The core has negligible effect on the strength of the fibers. The fibers are made by drawing down a rod or "preform" in a furnace. The fiber is immediately coated in-line by a liquid prepolymer which is then cured by ultraviolet radiation. The polymer coating protects the pristine as-drawn glass surface from subsequent handling damage, but is not hermetic and does not exclude moisture from the fiber surface. Although occasional defects can be found spaced apart by many meters, short length fiber specimens (less than ~1 m) exhibit a very narrow distribution of strength which is close to the theoretical value (14 GPa, 20% strain to failure). The fiber surface is therefore essentially flaw free[1] and strength is controlled by the surface condition rather than by defects inside the surface. Atomic force microscopy has proved an invaluable tool for directly observing the fiber surface morphology on the nanometer scale. This article describes three examples of where AFM has impacted our research into the strength and fatigue behavior of optical fibers.

FATIGUE-RESISTANT COATINGS

Fused silica glass, like all oxide materials, experiences slow weakening under the combined action of environmental moisture and applied stress. This phenomenon, known as fatigue or stress corrosion cracking, is caused by preferential attack by water of strained ≡Si–O–Si≡ bonds at the tip of a crack or other stress concentrator.[2] The bonds are ruptured to form two ≡Si–O–H groups and the crack front advances by one bond spacing. Under a constant applied load, this leads to slow growth of the crack until it reaches the critical size for catastrophic failure. The time to failure is a sensitive decreasing function of the applied stress; this phenomenon is known as delayed failure or, to ceramists, static fatigue. These processes can severely limit the use of oxide ceramics and other materials for applications of long duration in harsh environments.

In short term tests the static fatigue behavior of fused silica optical fiber appears qualitatively similar to that of weaker glasses even though the fiber does not contain sharp, well-defined cracks. For example, Figure 1 shows the static fatigue behavior for two different fibers tested in a harsh accelerated environment of 90° C, pH 7 buffer solution. For shorter times to failure, high applied stress (>2.5 GPa), the failure time increases rapidly as the applied stress decreases. However, at applied stresses less than 2.5 GPa there is a "knee" in the curve beyond which the failure time is much less sensitive to the applied stress and is substantially less than would be predicted from the short-term data. Fibers aged under zero applied stress in aggressive environments often show substantial strength loss at approximately the same time as the "knee" is observed in static fatigue and the two phenomena are therefore related.[3] Weaker materials do not exhibit such knee behavior; sometimes deviations are observed but they are in the opposite sense - *strengthening* upon severe aging is due to blunting of the crack tips and other processes. It is now known that the knees observed for high strength fiber in both static fatigue and zero stress aging are caused by the development of nanometer scale surface roughness; the surface pits are stress concentrators that become a new source of strength controlling defects.[4]

The relationship between roughness and strength has been demonstrated using both the STM[5] and AFM[6] techniques. Yuce et al[7] showed that while several different fibers showed quite different strength degradation kinetics in different environments, there was a single universal relationship between rms roughness measured using AFM and the residual strength.

Matthewson et al.[8] proposed that any mechanism that slows dissolution of the fiber surface should lead to an improvement in the strength behavior in harsh environments. They incorporated fumed silica particles (20 nm diameter) in the polymer coating of the fiber and found substantial improvements in the behavior. It is thought that these particles preferentially dissolve in moisture due to their high surface curvature, thus partially saturating the moisture with dissolution products makes them less reactive at the fiber surface. Figure 1 shows the static fatigue behavior of such fibers; the two fibers are identical except that the "with" fiber has 0.7 wt% of fumed silica incorporated in its polymer coating.[8,9] The powder substantially

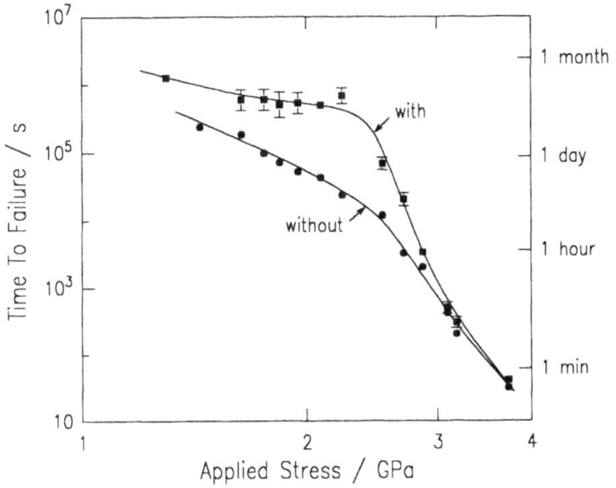

Figure 1. Static fatigue behavior in 90°C pH 7 buffer solution of two fused silica fibers which are identical except that the "with" fiber incorporates 0.7 wt% of fumed silica in the polymer coating.

delays the onset of the fatigue knee and failure times are up to 30 times longer beyond the knee. Higher concentrations of silica powder give proportionately more improvement and lifetime increases of up to 300 times have been observed for 3 wt% of additive.[10,11] Atomic force microscopy has been used to verify that the coating additive slows the formation of surface roughness.

Figure 2 shows AFM images of the fiber surface (after removal of the polymer coating) of (a) unaged fiber and fiber both (b) without and (c) with 0.7 wt% of silica powder in the coating that has been aged under zero stress for one week in 90°C, pH 7 buffer solution. The substantial roughening observed for the without fiber (Figure 1(b)) is suppressed by the silica additive (Figure 1(c)), corresponding to a significantly higher residual strength. The addition of reactive powders to the polymer coating to suppress strength degradation mechanisms is a simple modification of current coating technologies that promises to be an inexpensive method for fabricating fiber with substantially enhanced long-term mechanical reliability.

EFFECT OF COATING STRIPPING ON SURFACE MORPHOLOGY

While AFM has proved a valuable technique for determining the surface morphology of optical fibers, before it can be used the polymer coating must first be removed. The most common method for removing the coating before AFM analysis is to briefly dip the fiber in

200°C concentrated sulfuric acid. Other techniques that can be used are mechanical stripping and chemical stripping with, for example methylene chloride (MeCl) which swells the coating allowing easy removal by wiping. Both the latter techniques often result in abrasion of the fiber surface with resultant strength loss, and also can leave residual coating on the fiber surface. An important question is whether the stripping techniques change the surface morphology or the fiber.

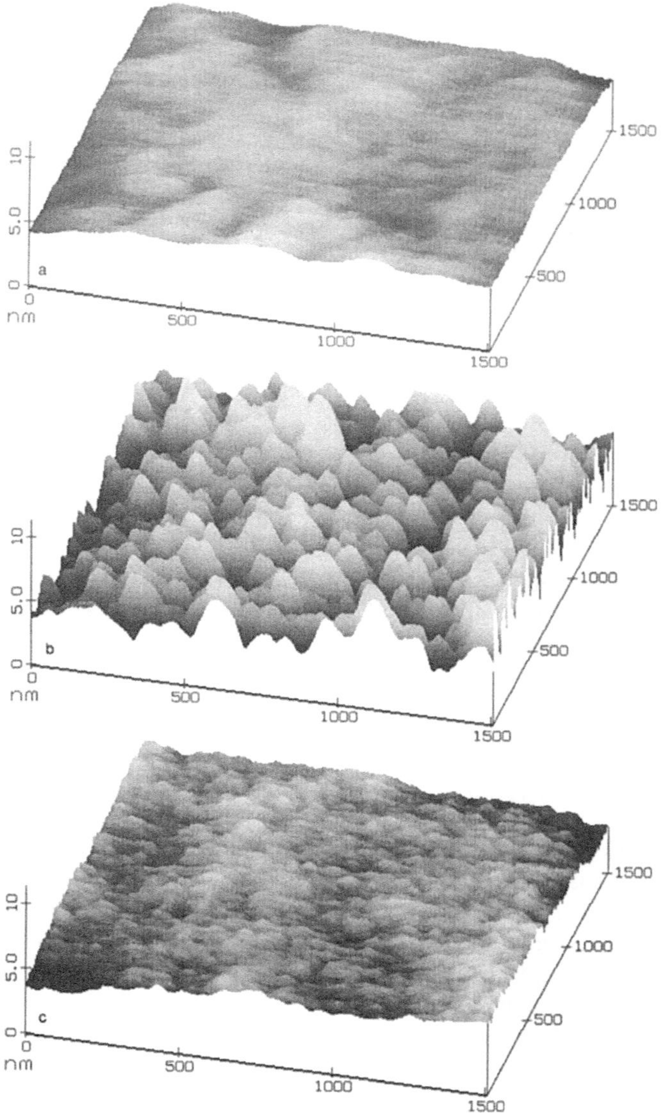

Figure 2. AFM images of the surface of (a) unaged fiber and fiber aged for 1 week in 90° C pH 7 butter solutions (b) without and (c) with silica powder additive in the polymer coating.

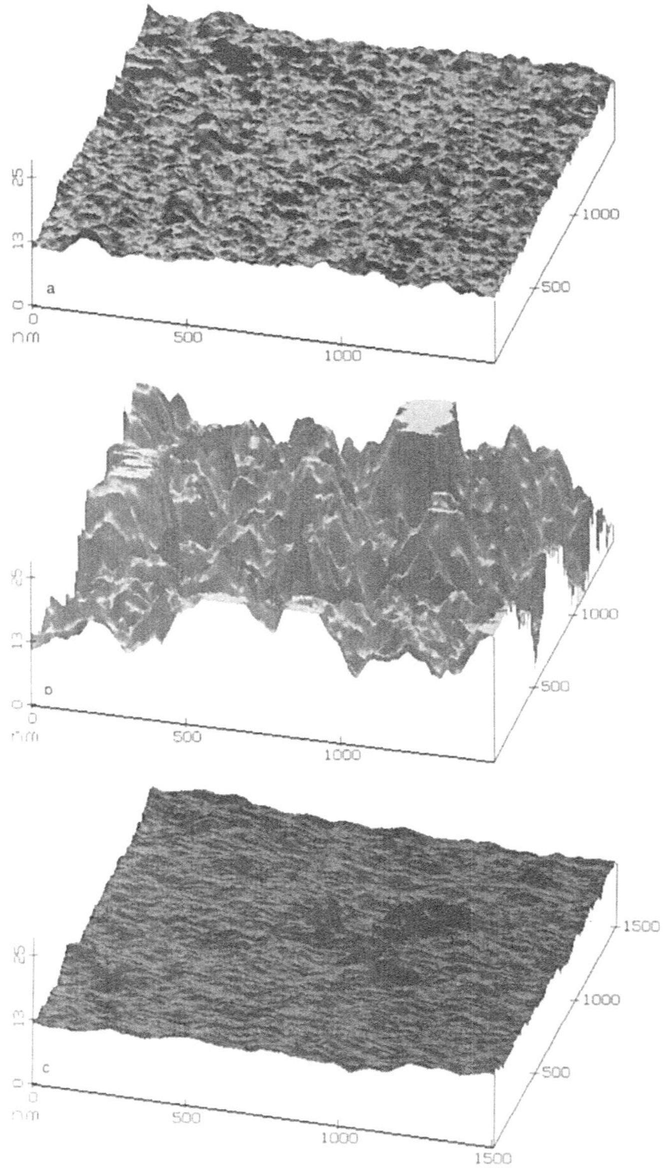

Figure 3. AFM images of the surface of silica fiber aged coated for 14 days in 90° C pH 7 buffer and subsequently stripped using (a) mechanical stripping, (b) hot acid immersion and (c) immersion in MeCl. Rms surface roughnesses are (a) 1.03 nm, (b) 6.3 nm and (c) 1.06 nm.

Measurements of the residual strength of fiber as a function of aging time show that the residual strength is the same whether the fiber is stripped using hot acid or MeCl.[12] This shows that while aging does substantially degrade the strength of the fiber, the measured strength is not affected by treating with either chemical. In contrast, the surface profiles determined by AFM differ substantially. Figure 3 shows the surface profiles of fiber aged for two weeks in 90°C, pH 7 buffer followed by stripping (a) mechanically, (b) in hot acid and (c) by MeCl. Except for the acid stripped surface, the rms roughnesses are approximately 1 nm, comparable to the roughness of the unaged surface of 0.5 nm. In contrast, the acid stripped surface is substantially rougher and corresponds more closely to the degraded strength of the aged fiber. These results indicate that a surface layer, probably of hydrated silica, is formed upon aging. This layer is not removed by either mechanical stripping or by MeCl immersion and is strong enough to support the AFM tip. However, the layer does not contribute to the strength of the fiber. The hot acid immersion does remove this layer, thus exposing the rougher strength controlling surface.

Figure 4 shows similar results for fiber that was aged bare, i.e., it was stripped in hot acid *before* aging. In this case the aged fiber was subsequently (a) untreated or dipped in (b) hot acid or (c) MeCl *after* aging. Again, the untreated and MeCl treated surfaces are similar to the unaged surface but the acid treated surface is substantially rougher. This shows that acid treatment must be used to expose the strength controlling surface even if there is no polymer coating to strip.

These results show that surface layers can form upon aging which, unless removed using some suitable process, can mask the underlying strength controlling surface. While the hot acid immersion does expose a rough surface that correlates with the residual strength, it does not prove that the true strength controlling surface has been exposed. Therefore, detailed quantitative analysis of AFM results may not be reliable unless it is clearly proved that the strength controlling surface and the surface imaged in AFM are one and the same.

HEAVY METAL FLUORIDE FIBERS

While fused silica glass is the material used for fabricating optical fibers for use in communications systems, other glasses may be used in other applications. For example, heavy metal fluoride fibers, often based upon zirconium fluoride, have transparency further into the infrared than silica and so have many applications such as for optical sensors or amplifiers. However, fluoride fibers are generally much weaker and less durable than silica.[13] AFM has proved useful in understanding the strength and strength degradation of fluoride fibers. Figure 5 shows the zero stress aging behavior, in 30°C pH 7 buffer solution, of two fluoride fibers, one based on zirconium fluoride ("ZBLAN") and one based on aluminum fluoride ("AlF").[14,15] The residual flexural strength of the fibers is determined as a function of the aging time. Throughout this work the residual strength is characterized by the strain to failure rather than the failure stress because this is what is directly measured by flexural techniques. Also, interpretation of the results does not require knowledge of the elastic modulus of the materials. In most fiber optic applications the fiber experiences a prescribed strain rather than stress so that the strain to failure is more relevant.

The ZBLAN fiber starts substantially weaker than fused silica and (given the low temperature of the test environment) also degrades much faster than silica. Figure 6 shows AFM images of the (a) unaged and (b) 100 min aged surfaces of ZBLAN fiber. The unaged surface shows some roughness which in part explains the relatively low initial strength compared with silica. The ZBLAN composition has a high rate of dissolution but has low solubility so that dissolution products quickly crystallize.[16] These crystals are clearly visible embedded in the fiber surface in Figure 6(b); they are constrained to form there by the presence of the polymer coating. However, the fiber's surface between the crystals is comparatively smooth and indeed is smoother than the unaged surface indicating that the dissolution leads to chemical polishing. In earlier work, Colaizzi et al.[17] found that fiber aged bare increased in strength upon aging in some test environments. In this case crystals are notformed at the fiber surface since there is no constraint by a coating; the removal of surface roughness is then responsible for the observed strength increase.

Figure 5 shows that, while the AlF fiber starts with a lower strength than the ZBLAN fiber, the rate of degradation is much slower leading eventually to a higher residual strength. This is because the AlF composition is much less soluble in water than the ZBLAN composition. Figure 7 shows AFM images of the unaged and aged (10^4 min) surfaces of the AlF fiber which show little change in the morphology upon aging. These results show that if fiber is to be exposed to harsh environments, the initial strength is of little value in predicting long-term durability since the degradation rate will usually dominate.

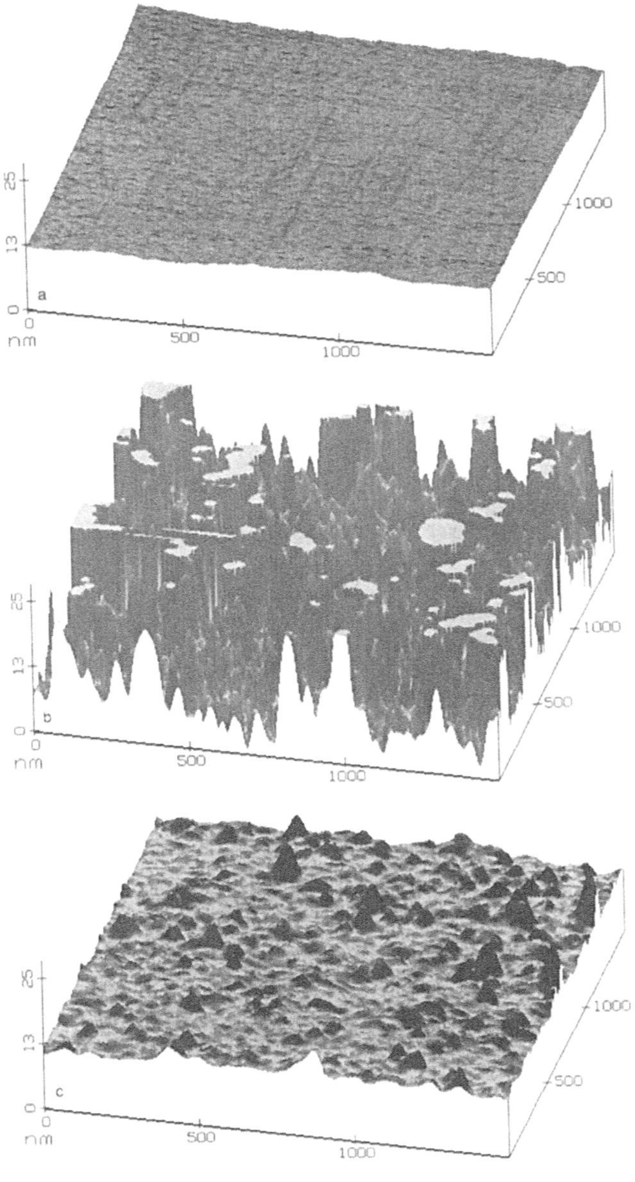

Figure 4. AFM images of the surface of silica fiber aged bare for 14 days in 90° C pH 7 buffer and subsequently (a) untreated, (b) dipped in hot acid and (c) immersed in MeCl. Rms surface roughnesses are (a) 0.56 nm, (b) 8.3 nm and (c) 0.85 nm.

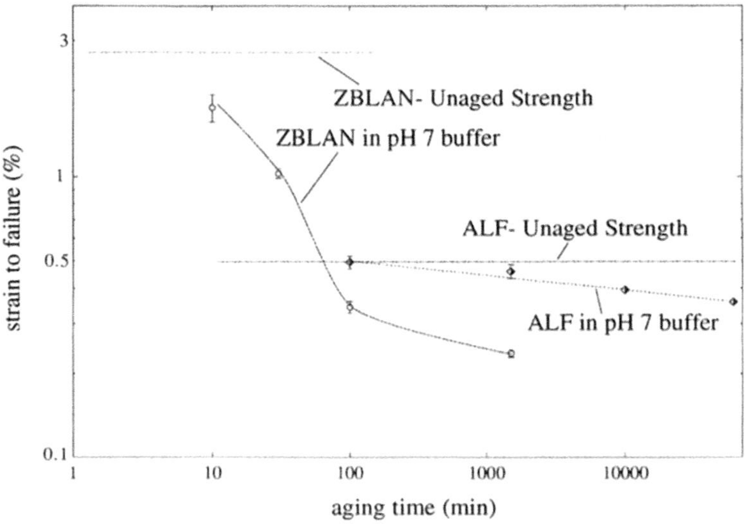

Figure 5. Strength of ZBLAN and AlF fluoride fibers as a function of aging time in 30° C pH 7 buffer solution.

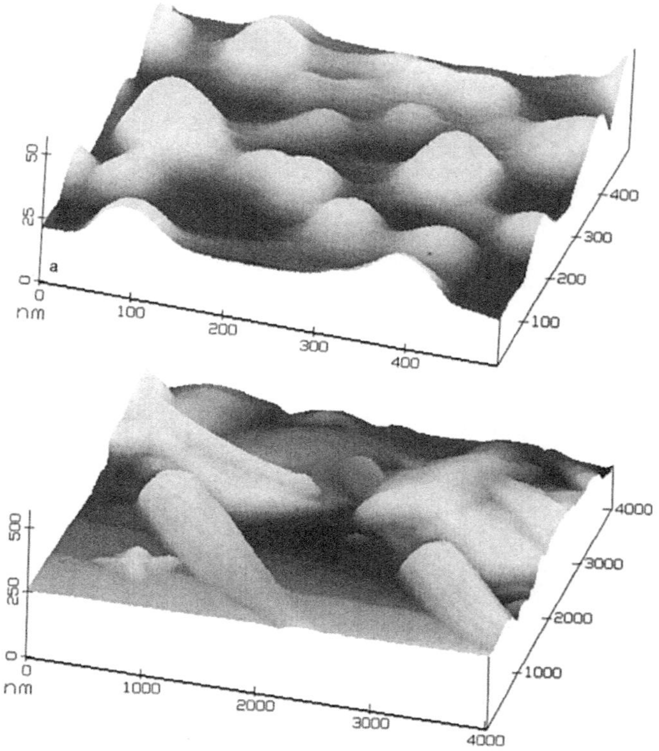

Figure 6. AFM images of the surface of ZBLAN fiber (a) unaged and (b) after aging for 100 min in 30°C pH 7 buffer solution (note the different vertical resolutions).

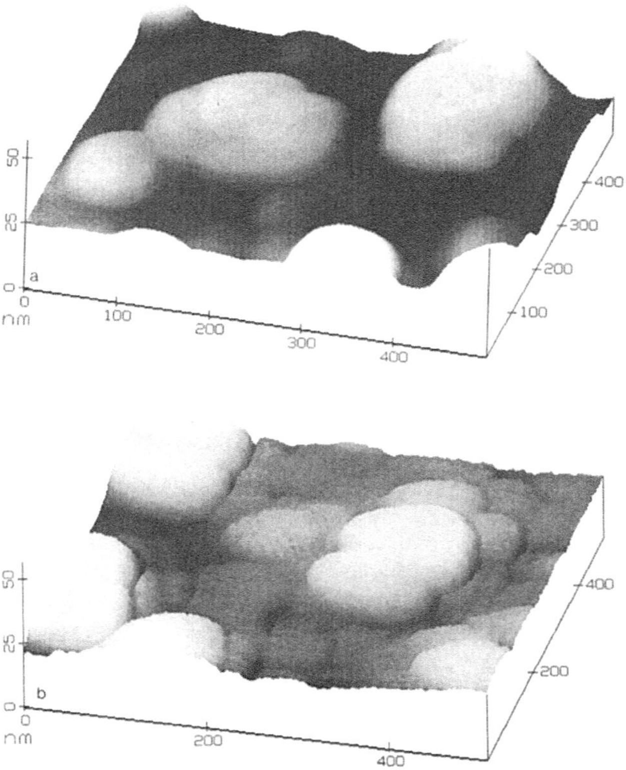

Figure 7. AFM images of the surface of AlF fiber (a) unaged and (b) after aging for 10^4 min in 30°C pH 7 buffer solution (note the different horizontal scales).

CONCLUSIONS

Compared to other ceramic materials optical fibers are very strong. However, their use in aggressive environments results in strength degradation that can lead to eventual mechanical failure. The degradation is caused by the formation of surface roughness due to dissolution of the fiber material in environmental moisture. While the roughness may only be a few nm rms, it is sufficient to severely degrade the strength.

AFM has proved a valuable tool for imaging and quantifying the surface roughness. Its importance in understanding and predicting the mechanical reliability of optical fibers is expected to increase.[18]

REFERENCES

1. C.R. Kurkjian and U.C. Paek, Single-valued strength of 'perfect' silica fibers, *Appl. Phys. Lett.*, 42: 251-253 (1983).
2. T.A. Michalske and B.C. Bunker, A chemical kinetics model for glass fracture, *J. Am. Ceram. Soc.*, 76: 2613-2618 (1993).
3. M.J. Matthewson and C.R. Kurkjian, Environmental effects on the static fatigue of silica optical fiber, *J. Am. Ceram. Soc.*, 71: 177-183 (1988).

4. C.R. Kurkjian, J.T. Krause and U.C. Paek, Tensile strength characteristics of 'perfect' silica fibers, *J. de Phys.*, 43: C9-585-586 (1982).
5. R.S. Robinson and H.H. Yuce, "Scanning tunneling microscopy study of optical fiber corrosion: surface roughness contribution to zero-stress aging," *J. Am. Ceram. Soc.*, 74: 814-818 (1991).
6. H.H. Yuce, J.P. Varachi, Jr. and P.L. Key, "Effect of zero-stress aging on mechanical characteristics of optical fibers," *Ceram. Trans.*, 20: 191-203 (1991).
7. H.H. Yuce, J.P. Varachi, Jr., J.P. Kilmer, C.R. Kurkjian and M.J. Matthewson, "Optical fiber corrosion: coating contribution to zero-stress aging," *OFC'92 Tech. Digest*, Postdeadline paper-PD21 (1992).
8. M.J. Matthewson, V.V. Rondinella and C.R. Kurkjian, "The influence of solubility on the reliability of optical fiber," *Proc. Soc. Photo-Opt. Instrum. Eng.*, 1791: 52-60 (1992).
9. V.V. Rondinella, M.J. Matthewson and C.R. Kurkjian, "Coating additives for improved mechanical reliability of optical fiber," *J. Am. Ceram. Soc.*, 77: 73-80 (1994).
10. M.J. Matthewson, H.H. Yuce, V.V. Rondinella, P.R. Foy and J.R. Hamblin, "Effect of silica particles in the polymer coating on the fatigue and aging behavior of fused silica optical fiber," *OFC'93 Tech. Digest*, 4 PD21 87-90 (1993).
11. V.V. Rondinella, M.J. Matthewson, P.R. Foy, S.R. Schmid and V. Krongauz, "Enhanced fatigue and aging resistance using reactive powders in the optical fiber buffer coating," *Proc. Soc. Photo-Opt. Instrum. Eng.*, 2074: 46-51 (1993).
12. V.V. Rondinella and M.J. Matthewson, "Effect of chemical stripping on the strength and surface morphology of fused silica optical fiber," *Proc. Soc. Photo-Opt. Instrum. Eng.*, 2074: 52-58 (1993).
13. S.F. Carter, "Mechanical properties" in "Fluoride glass optical fiber," eds. P.W. France, M.G. Drexhage, J.M. Parker, M.W. Moore, S.F. Carter and J.V. Wright, 219-237, CRC Press, Boca Ratan FL (1990).
14. J.J. Colaizzi, M.J. Matthewson, T. Iqbal and M.R. Shahriari, "Mechanical properties of aluminum fluoride glass fibers," *Proc. Soc. Photo-Opt. Instrum. Eng.*, 1591: 26-33 (1991).
15. J.J. Colaizzi and M.J. Matthewson, "Mechanical durability of ZBLAN and aluminum fluoride-based optical fiber," *J. Lightwave Tech.*, 12: 1317-1324 (1994).
16. D.G. Chen, C.J. Simmons and J.H. Simmons, Corrosion layer formation of ZrF_4-based fluoride glasses, *Mat. Sci. For.*, 19-20: 315-320 (1987).
17. J.J. Colaizzi, M.J. Matthewson, M.R. Shahriari and T. Iqbal, "Environmental effects on the mechanical properties of aluminum fluoride glass optical fibers," *Ceram. Trans.*, 28: 579-586 1992.
18. M.J. Matthewson, "Optical fiber reliability models,". *Soc. Photo-Opt. Instrum. Eng. critical review series*, CR50 32-59 (1994).

SCANNING TUNNELING MICROSCOPY STUDIES OF SOLVENT-DEPOSITED MATERIALS ON HIGHLY ORIENTED PYROLYTIC GRAPHITE

Edwin J. Hippo and Deepak Tandon
Department of Mechanical Engineering and Energy Processes
Southern Illinois University at Carbondale
Carbondale, Illinois 62901

Abstract: Scanning Tunneling Microscope (STM) is a new tool for resolving surface topography and electronic states at the atomic level. STM excels over other microscopy techniques in its high lateral and vertical resolutions. Many forms of carbon including diamonds, fullerenes, isotropic porous carbons and glassy carbons have been studied extensively by STM. STM studies of basic carbon materials and surface adsorbed species on these carbon surfaces are essential as a foundation for future nanoscopic work involving more complex materials, such as carbon-carbon composites and coal structures, as well as the advancement of carbon inhibition/gasification technology. Our studies have shown that atomic structure of solvent deposited coal can be revealed and this will help in better understanding of the coal structure. The metal salts deposited on the graphite structure will help in a better understanding of the catalysis in carbon gasification. In this paper we have discussed some of the preliminary work performed on the materials deposited on the highly oriented pyrolytic graphite (HOPG) surface.

INTRODUCTION

Scanning tunneling microscope (STM) is a new tool for resolving surface topography and electronic states at the atomic level. STM excels over other microscopy techniques in its high lateral and vertical resolutions. Tunneling occurs when the wave functions of a sharp tip and the sample surface overlap. The sample surface has to be conducting or semiconducting. If a positive voltage (bias voltage) is applied than the electrons tunnel from the tip to the sample surface, a negative voltage results in tunneling from the surface to the tip. Many forms of carbon including diamonds,[1-3] fullerenes,[4,5] isotropic porous carbons,[6-8] and glassy carbons[9] have been studied extensively by STM. However, it is the study of highly oriented pyrolytic graphite that demonstrates the STM's ability to resolve atomic images of carbon materials. The (001) basal plane of HOPG has a very regular structure, it is conductive and it is chemically inert under tunneling conditions. These factors allow individual atoms at the surface to be imaged under atmospheric conditions and account for the use of HOPG as a STM calibration material.[9-11] Typically the 142 pm carbon-carbon bond length and the 250 pm hexagonal ring center-to-center distance[9-11] observed for the HOPG basal plane, and the 335 pm spacing between basal planes[12] are used to calibrate the instrument.

STM studies of basic carbon materials and surface adsorbed species on these carbon surfaces are essential as a foundation for future nanoscopic work involving more complex materials, such as carbon-carbon composites and coal structures, as well as the advancement of carbon inhibition/gasification technology. One area of coal research that is not clearly understood is the modeling of coal structures. STM may be used to elucidate coal structure on an atomic level. We have imaged dissolved coal components deposited on HOPG by STM. We have also imaged n-dotriacontane (a long-chain carbon molecule), and metal salts e.g., sodium and cesium chloride deposited on the surface of HOPG. Our studies have shown that atomic structure of solvent-deposited coal can be revealed and this will help in better understanding of coal structure and would possibly enhance sulfur/mineral impurity beneficiation methods. Metal salt images will help in better understanding the role of catalysts during carbon gasification reactions.

In this paper we have discussed some of the preliminary work performed on the materials deposited on the HOPG surface.

EXPERIMENT

All STM nanographs were obtained using a Nanoscope II scanning tunneling microscope manufactured by Digital Instruments of Santa Barbara, CA. Pristine samples of HOPG were obtained from Union Carbide. For all the images Pt/Ir (80/20) tips manufactured by Digital Instruments were used. The distances were measured to within ± 10 pm of the theoretical distances. Bias voltages of 13.1 and 20.14 mV and a current of 1.99 nA were used. All images were obtained in the constant current mode. Sodium Chloride, Cesium Chloride, n-dotriacontane, and THF extracts of Herrin no. 6 coal were deposited by solvent deposition technique.

RESULTS AND DISCUSSION

Graphene layers in HOPG are staggered in such a way that every other carbon atom is directly over another carbon atom in the graphene layer below.[10] It is at these points of direct carbon atom overlap where basal planes are held together by weakly attractive van der Waals forces. Calculations predict that at these points there is an increased overlap of the electronic clouds in such a way that the electronic clouds of these atoms are slightly smaller and therefore lower than the electronic clouds of the surface carbon atoms that do not have another carbon atom directly beneath and thus every other carbon atom should be drawn slightly closer together. Since this occurs for every other carbon atom the surface of HOPG should show a slight corrugation due to the alternating attracted and unattracted carbon atoms at the surface. According to the Soler et al. model[12] the alternate depressed clouds are not imaged by the STM tip and a trigonally symmetric structure is observed. This argument leads to an interpretation of HOPG basal plane images in which only every other carbon atom protrudes from the surface - the three-carbon' atom model. We have imaged HOPG basal plane and have consistently imaged all six carbon atoms in the hexagonal ring structure of the basal plane.

Figure 1(a) is a 2 x 2 nm top-view image of a HOPG basal plane. The distance between adjacent yellow spots is 250 pm which is the expected distance for a ring center to ring center measurement. The distance between the dark red and light red/orange is 142±10 pm, which is the theoretical carbon-carbon bond length. Figure 1(b) is an enlarged portion of the image in Figure 1(a). The image is plotted in pseudo 3-D mode at a pitch angle of 30° to obtain clearer nanograph that confirms the presence of all six-carbon atoms in the hexagonal ring structure.

We have imaged HOPG basal planes at different bias voltages and have observed the center of the rings appear both as bumps and pits. This could be because of tunneling through different orbitals. When tunneling occurs between the orbitals bumps are observed and when the tunneling occurs between the sigma orbitals pits are observed in the center of the rings.

Figure 2(a) is a 2.2 x 2.2 nm top view image of solvent deposited Herrin, IL no. 6 coal THF extract. The image reveals the atomic structure of the deposited coal, with several groups of benzene rings evident, as well as side-chains branching off from these rings. Figure 2(b) is a 3 x 3 nm top view image of solvent deposited Herrin, IL no. 6 coal THF

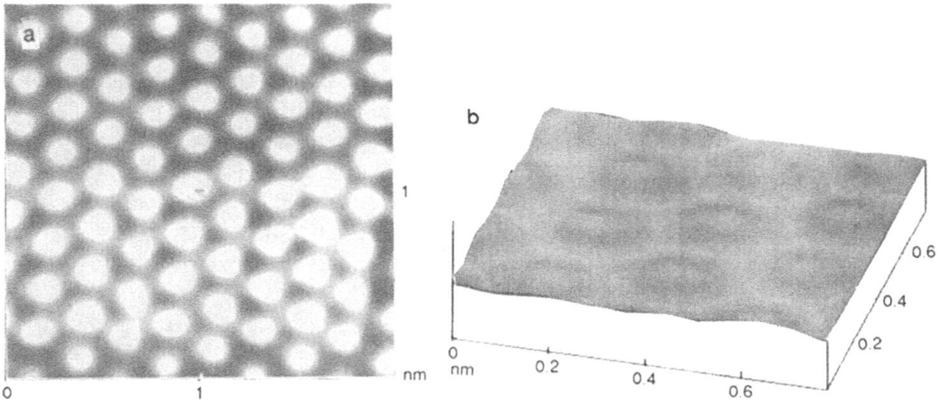

Figure 1. (a) Top view of HOPG basal plane. Trigonal symmetry with a hexagon. (b) Surface view of an enlarged portion of image in Figure 1(a), showing clear hexagonal ring.

extract. The middle portion of the image shows heavier coverage by the coal, and reveals how the coal molecules align along the carbon atoms in the basal plane of HOPG. The bottom portion of the image suggests some atomic ring structures with side chains.

Figure 3(a) is a 2 x 2 nm top view image of solvent deposited n-dotriacontane (n-C_{25}) on HOPG. It was envisioned that if this typical long chain carbon molecule could be imaged, others such as those in coals and pharmaceutical products might be imageable as well.

The image seen in Figure 3(a) depicts the n-C_{25} on the surface of HOPG. Spacings of 0.3 nm, instead of 0.25 nm, between hexagon centers, as well as the elliptical shape of the molecules suggest the presence of n-C_{25}. The molecules appear elliptical, due to the hydrogen atoms attached to the carbon atoms. The somewhat repeating pattern on the

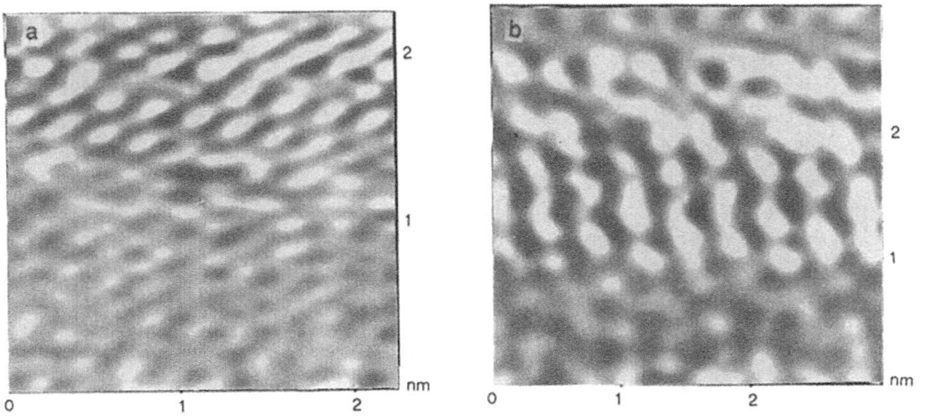

Figure 2. (a) Top view of THF extract of IL no. 6 coal on HOPG surface, groups of benzene rings are evident. (b) Top view of THF extract of IL no. 6 coal on HOPG surface, with heavy coverage in the middle portion of the image.

surface is due to the fact that the n-C_{25} molecules align themselves along the carbon atoms of the top basal plane layer of HOPG. Figure 3(b) is a 2 x 2 nm top view image of solvent deposited n-C_{25} on HOPG basal plane. The coverage of the HOPG basal plane by n_{25}C molecules is heavy. The molecules have aligned along the rows of the carbon atoms on the HOPG surface and once again the spacing between the centers has increased to 0.3 nm.

Figure 4(a) is a 2 x 2 nm 3-D surface image of sodium chloride solvent deposited on the surface of HOPG. The image shows the hexagonal up and down pattern because the size of the chloride ion (ionic radii 0.18 nm) is almost double the size of sodium ion (ionic

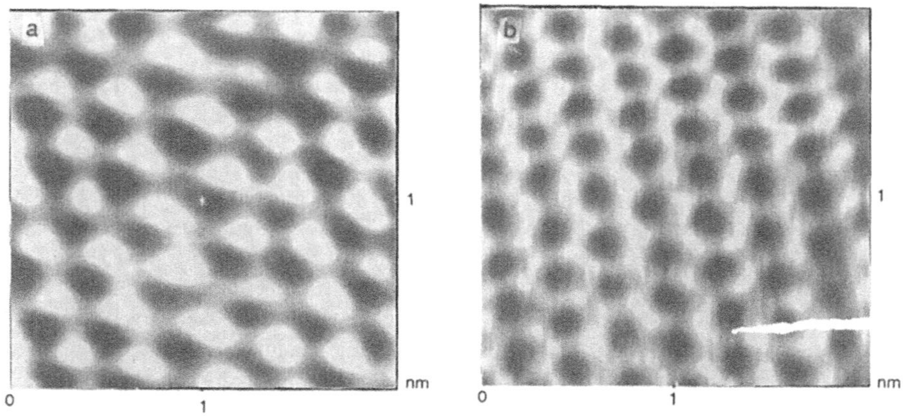

Figure 3. (a) Top view of solvent deposited n-dotriacontane on HOPG, larger spacing confirms the presence of n-C_{25}. (b) Top view of solvent deposited n-dotriacontane on HOPG, with heavier coverage of n-C_{25}.

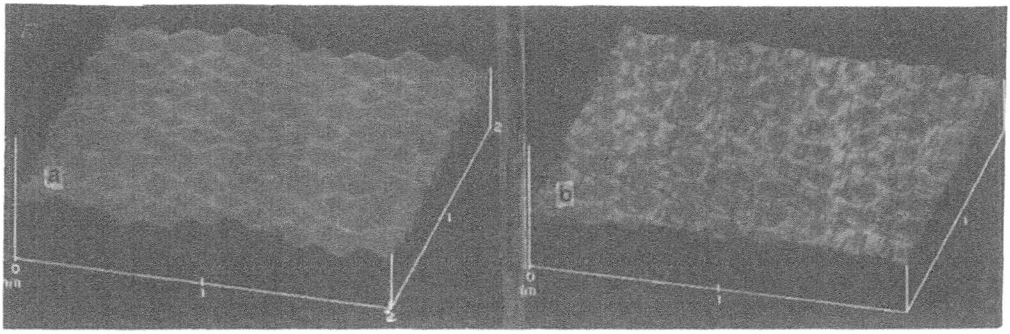

Figure 4. (a) Surface view of sodium chloride deposited on the surface of HOPG. The up and down pattern of the hexagon shows sodium and chloride ions on top of alternate carbon atoms. (b) Surface view of cesium chloride deposited on the surface of HOPG. No up and down pattern of the hexagon is observed because the cesium and chloride ions have about the same ionic radii.

radii 0.097 nm). Sodium and chloride ions sit alternately on top of the carbon atoms in the basal plane of HOPG.

Figure 4(b) is a 2 x 2 nm 3-D surface image of cesium chloride solvent deposited on the surface of the HOPG. The image shows the hexagonal raised structures and there is no up and down topography as was the case with sodium chloride, this is because the ionic radii of chloride and cesium (0.167 nm) ions are comparable. One interesting feature that we observed was that the salts cannot be washed with water after they were deposited.

CONCLUSIONS

All six carbon atoms of the hexagon in HOPG basal plane have been imaged. The correct interpretation of the basic structure is important in order to study more complex structures. Nonconductive structures can be studied by depositing them on a conductive surface. We have imaged metal salts, coal extracts, and long chain hydrocarbons. These species can be observed by varying the bias voltage which in turn results in tunneling through different orbitals. These studies will help in a better understanding of the coal structure and role of catalyst during gasification.

REFERENCES

1. H.G. Busmann, H. Sprang, I.V. Hertel, W. Zimmermann-Edling, and H.J. Guntherodt, Scanning Tunneling Microscopy on Chemical Vapor Deposited Films, *Appl. Phys. Letters*. 59: 295-297 (1991).
2. T. Tsuno, T. Imai, Y. Nishibayashi, K. Hamada, and N. Fujimori, Epitaxially grown dimond (001) 2 x 1/1 x 2 surface investigated by scanning tunneling microscopy in air, *Jpn. Journal of Appl. Phy.* 30: 1063-1066 (1991).
3. W. Zimmermann-Edling, H.G. Busmann, H. Sprang, and I.V. Hertel, *Ultramicroscopy*, 42-44: 1366 (1992).
4. R.J. Wilson, G. Meijer, D.S. Bethune, R.D. Johnson, D. D. Chamblis, M.S. de Vries, H.E. Hunziker, and H.R. Wendt, Imaging C_{60} Clusters on a Surface Using a Scanning Tunneling Microscope, *Nature*, 348: 621-622 (1990).
5. J.L. Wragg, J.E. Chamberlin, H.W. White, W. Kratschmer, and D.R. Huffman, Scanning Tunneling Microscopy of Solid C_{60}/C_{70}, *Nature*, 348: 623-624 (1990).

6. N.H. Cho, D.K. Veirs, J.W. Ager III, M.D. Rubin, C.B. Hopper, and D.B. Bogy, "Effects of Substrate Temperature on Chemical Structres of Amorphous Carbon Films." *J. Appl. Phys.*, 71: 2243-2248 (1992).
7. W.P. Huffman, V.B. Elings, and J.A. Gurley, Scanning Tunneling Microscopy of Carbon Fibers, *Carbon*, 26: 754 (1988).
8. H.N. Lei, A. Metrot, and M. Troyon, Scanning Tunneling Microscopy Analysis of a Glassy Carbon Heat Treated at 2500°C, *Carbon,* 32: 79-84 (1994).
9. S. Gauthier et al. A Study of Graphite and Intercalated Graphite by Scanning Tunneling Microscopy, *J. Vac. Sci. Technol.*, A6: 360-362, (1988).
10. I.P. Batra and S. Ciraci, Theoretical Scanning Tunneling Microscopy and Atomic Force Microscopy Study of Graphite Including Tip-Surface Interaction, *J. Vac. Sci. Technol.* A6:313-318, (1988).
11. L.L. Soethout, Random Packing and Random Tessellation in Relation to the Dimension of Spacing, *J. Microscopy*, 152-1:251-258, (1988).
12. J.M. Soler, A.M. Baro, N. Garcia, and H. Rohrer, Interatomic Forces in Scanning Tunneling Microscopy: Giant Corrugations of the Graphite Surface, *Phys. Rev. Lett.,* 57:(4); 444-447 (1986).

IN SITU STUDY OF STAINLESS STEEL'S PASSIVE LAYER EXPOSED TO HCL USING A SCANNING TUNNELING MICROSCOPE

T. J. McKrell and J. M. Galligan

University of Connecticut
Institute of Materials Science
Department of Metallurgy
Storrs, CT 06269

Abstract: The nature of 304 stainless steel's passive film and how it is altered as a function of exposure time in hydrochloric acid (HCl) acid has been studied using a scanning tunneling microscope (STM). This experiment offers the distinct advantage of *in situ* studies, on a microscopic scale, of the electrical character of the surface in the presence of a corrosive environment. In the present paper we present evidence which shows directly how the surface structure of 304 stainless steel is altered by HCl; simultaneously the surface is probed with a STM. In particular, the *in situ* current, I, versus bias voltage, V, curves were measured as a function of etchant exposure time. The important region, concerning passivation, is the extent and the slope of the I-V curve near where the current is zero; this region corresponds to the material showing corrosion resistance and exhibits a low or zero slope near where the current is equal to zero. The research establishes a quantitative measure of the extent of corrosion protection, stability, and the kinetics of the passive film. The findings of this research are correlated with nanoscale STM images of the 304 stainless steel's surface.

INTRODUCTION

Passivity is an unusual phenomenon observed during the corrosion of certain metals and alloys. Simply, it can be defined as a loss of chemical reactivity under certain environmental conditions.[1] Passivation is not a constant state or an inherent property, but may only apply to certain environments or conditions and even when a metal or alloy is passive under a given set of conditions a slight change in these conditions may destroy the protective film and so change the metal or alloy from the passive to the active state.[2] This process, needless to say has been studied extensively for a very long time, and yet there are still major questions that are unanswered. For example, what is the nature of the passive film and how is this altered when such surfaces are present in corrosive environments.

One way of getting some insight into this problem is by examining the surface with a scanning tunneling microscope (STM). This device offers the distinct advantage of *in situ* studies of the electrical character of the surface in the presence of a corrosive environment, with nanometer resolution. Experiments described below using the STM provide new ways of examining basic questions about the nature of passivity.

In the present paper we present evidence that shows how the extent of corrosion protection and stability of 304 stainless steel's passive film is altered by hydrochloric acid (HCl). That is while the surface is attacked by HCl the surface's electrical character is established with an STM in a spectroscopic mode. In particular, the *in situ* current, I, and bias voltage, V, curves were measured in order to determine how these varied with etchant exposure time. The kinetics of the passive film were also investigated using the STM in an imaging mode. An applied potential was observed to result in the growth of the passive film which has been quantitatively measured with the STM in an imaging mode. These experiments are described below.

BACKGROUND MATERIAL

In spite of the intensive effort, relatively little is known of the electronic and physical properties of passive films.[3] Several models have been proposed to explain the electronic and physical properties of passive films, so as to understand their growth and dissolution. These models suggest that passive films are highly doped semiconductors, existing under the influence of a strong electric field. Experimentally, it has been found that film growth kinetics follow one of the following laws[4]

$$L = A + B\ln t \quad \text{(logarithmic)} \tag{1}$$

$$1/L = C - D\ln t \quad \text{(inverse logarithmic)} \tag{2}$$

where:
L = film thickness
t = time
A, B, C, D = constants dependent upon material properties and the applied voltage between the corrosive medium and the metal.

Further, it is generally found that the dissolution of the passive film and subsequent corrosion is a function of:[5]

1. The potential across the film exceeding some critical value referred to as the breakdown potential. This breakdown potential is a function of halide ion activity according to

$$V_c = E - F\ln a_{x^-} \tag{3}$$

where:
E,F = constants
a_{x^-} = activity of the halide ion.

2. The dissolution of the passive film occurs at highly localized sites such that this localized dissolution ultimately results in the formation of a pit. The time required from the addition of the halide to the solution until the nucleation of the pit is referred to as the incubation time.

3. Ellipsometric work by McBee and Kruger[6] reveals that during the incubation period passive films undergo a sudden change, which might be attributed to the incorporation of Cl^- anions into the passive film.

There are a number of models for the dissolution of passive films. These are summarized in reference 7. The point defect model, given by C.Y. Chao, L.F. Lin, and D.D. Macdonald, appears to be the most consistent with the known facts about corrosion. The point defect model is based upon the movement of defects in an electrostatic field. This model states that the dissolution of the metal results from the flux of cations to the film/solution interface leaving behind metal vacancies or metal holes at the metal/film interface. These metal "holes" will tend to submerge into the bulk of the metal and hence to "disappear".[7] However, if the production rate of these holes is larger than their submergence rate, then the holes will pile up at the metal surface, forming a void. As the void grows to a critical size the passive film collapses down into the void resulting in locally accelerated corrosion and hence the formation of a pit. The presence of the halide (Cl^-) is the catalyst that causes the production rate of the holes to surpass the submergence rate, and therefore the presence of the halide determines if chemical breakdown and pit initiation will occur. A closely related model is presented below.

The model proposed in this paper incorporates the influence of the structural aspects of the base metal and passive film upon the dissolution of the metal and passive film. For example, point defects, the products of corrosion, are charged and these charged entities interact with dislocations and grain boundaries of the passive film. Further, these dislocations and grain boundaries offer short circuits for these defects to diffuse into the bulk metal and out into solution. These short circuits locally accelerate the dissolution of the metal surface leading to the formation of pits. Also, dislocations and grain boundaries within the passive film affect the thickness of the passive film, because the charge associated with dislocations in semiconductors would have an effect on the electric field over the thickness of the passive film, equations 1 and 2. The presence of charged dislocations and grain boundaries within the passive film would, therefore, lead to non-uniformity of the passive film's thickness. Such nonuniformity would result in locally thinner areas making the material more susceptible to corrosion, because diffusion lengths are effectively shorter, further adding to the chance of pit initiation. Further, the heterogeneous character of such defects would substantially reduce the nucleation barriers for the formation of a pit.

This model, like the point defect model, considers the passive film to contain a large number of defects and that growth and dissolution is a result of the creation and movement of these defects. The present model, unlike the point defect model, establishes a basis for the mechanisms involved in the nucleation and growth of pits, and, thus the corrosion processes. A somewhat more complicated scenario would be required for so-called amorphous films, but the idea of charged entities interacting may still apply.

EXPERIMENTAL AND RESULTS

Experiments to measure the I-V curves were carried out as follows: The apparatus consisted of a STM and a sample fluid cell. The surface of the 304 samples (0.019" thick by 0.5" in diameter) was mechanically polished to a mirror finish then rinsed with methanol and allowed to passivate in air. The sample fluid cell was constructed by attaching an o-ring to the sample's surface. In this manner a cell was formed into which 20% HCl was injected. The fluid was introduced with the STM engaged to ensure that the leakage currents between the corrosive solution and tip were not large enough to cause a loss of image.

An iridium-platinum tip was insulated with a glue and then microstop, in order to minimize the leakage currents; this enhanced the accuracy of the measured I-V curves. The tip to sample bias voltage, V_b, setpoint current, I_s, and instrument gains were held constant for all measurements of the I-V curves. The measured I-V curve of air passivated stainless steel, obtained as described above, is shown in Figure 1; the curve comprises three distinct regions–two active and one passive region, again Figure 1. Figure 2 illustrates how the I-V curves change as a function of exposure time to the acid.

The kinetics of the alteration of the passive film thickness as a function of applied potential between the tip and sample were measured using the STM in an image mode. This study was conducted under atmospheric conditions. The tip was spatially fixed over the center of the stainless steel's surface while the applied potential was held at 4V. STM images were then taken sequentially after the removal of the applied potential and the morphology changes were determined as a function of time.

The stainless steel's morphology changes, resulting from the application and subsequent removal of a spatially fixed potential, are shown in Figure 3. Finally Figure 4 shows how the diameter of the oxide growth changes with time after the potential is removed. Note that when measuring these changes in morphology a substantially lessened field is used. Also, the time that the reduced field is applied to a given area is greatly less when measuring the morphology.

DISCUSSION

The results illustrated above show the following:

1. Passive and active regions are present in the air passivated stainless steel, Figure 1.
2. Addition of HCl to the fluid cell shows that subsequent changes in the passive film take place on the stainless steel; that is the removal is nonuniform, consistent with the model proposed, Figure 2.
3. Quantitative measurements of the passive film's dissolution are given, Figure 2.
4. The morphology measurements also show that the passive film is nonuniform, Figure 3.

These results demonstrate a nonuniform film on the surface of the material. Given that the entities that are responsible for this change in passive film thickness, and the introduction of pits, are charged then these charged entities will naturally be attracted to dislocations and small angle boundaries, which are also charged in semiconductors.[8] Such a situation naturally leads to two things,

a. a nonuniform removal of the material, as observed, and

b. a nonuniform distribution of the electric field within the oxide such that preferential removal of material will take place at such charged sites. Again these observations are consistent with the model described above. This is elaborated below.

The observed I-V curve (Figure 1) is typical, for example, of a semiconducting material. It shows a very low if not a zero slope near V=0. Also, this region corresponds to a corrosion resistant, passive film. The passive region is of particular significance because information about the corrosion process can be determined from its slope and its width. Since the proposed model states that the dissolution of the passive film results from the movement of point defects in an electrostatic field, then this motion would be enhanced by the charged character of the defects involved. Further, a zero slope, shows that for the applied potential used and the defects present it is insufficient to allow charged entities to surmount a barrier. Accordingly, the slope of this region shows that dissolution, that is corrosion, is not taking place. Also, the passive region is bounded by the breakdown potentials, again Figure 1. These breakdown potentials correspond to the energy required to have a charge jump over an energy barrier. At the breakdown potential and beyond the tunneling current is given as[9]

$$I = GV_b\exp(-Hd) \qquad (4)$$

where:

G = proportionality constant

V_b = bias voltage (tip-to-sample)

H = proportional to the square root of the energy barrier height between the sample and tip

d = tip to sample distance.

Consider now how the tunneling current relates to acid exposure. Measurements of the change in the I-V curves as a function of etchant exposure time are shown in Figure 2. These curves show that with increasing time the width of the passive region decreases and at the same time the slopes of the passive and active regions increase. These changes can be explained by the introduction of and the motion of charged defects. The energy barrier height, for example, would be expected to decrease with time due to the presence of charged defects especially if these charged entities provide gap states in a semiconductor, eventually leading to a metallic state. That is, as more and more charged particles are added to the semiconductor, then the material will become a so-called degenerate semiconductor. This provides short circuits paths with pronounced corrosion occurring at these sites. But the width change and barrier height reduction are directly related, and so the width can be used as a measure of the stability of the passive film under the influence of various potentials and concentrations. For example, a decreasing width shows that the stability of the passive film is decreasing with increasing etchant exposure time. In addition the

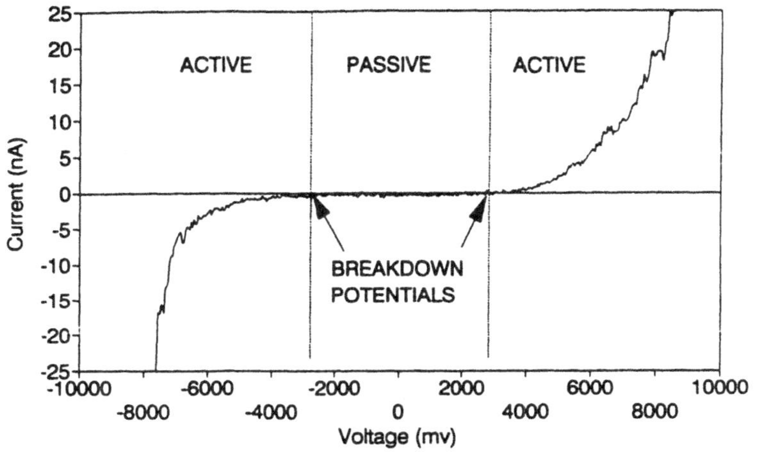

Figure 1. Typical I-V curve of air passivated 304 stainless steel measured with the STM in spectroscopic mode. Labeled are the active and passive regions along with the breakdown potentials.

slope of the I-V curve corresponds to a change in current resulting from a change in potential. An increasing slope with time shows that a larger change in current results from a constant change in applied potential. This increasing current with time indicates that the energy barrier height is also decreasing with time, Eq. 4. A decreasing energy barrier height with time, like the reduction in passivity, is due to the increasing number of charged defects formed as a result of the corrosion process.

The proposed model suggests, then, that this flow of charge is directly responsible for corrosion. Accordingly, the increasing slope with etchant exposure time, Figure 2, shows that the extent of corrosion protection provided by the passive film deteriorates with time, as expected. This is also consistent with the model presented. Further it goes beyond previous models, in that it shows how the film is expected to breakdown with time.

Consider now the morphology of the film. The morphology changes, as a consequence of an applied potential between the tip and sample, shown in Figure 3 are consistent with the passive film thickness varying logarithmically with time, as shown in Figure 4. The surface topology images, shown in Figure 3 are sequentially arranged with the upper left frame showing the initial state of the 304 passive film. Between the first and second frames the tip was held at the center of the sample at 4V resulting in a localized thickening of the passive film, which is evident in the second frame. The remaining frames show the sequential decay of this passive film growth after the potential was removed. The decay of the passive film growth, shown in Figure 4, was obtained

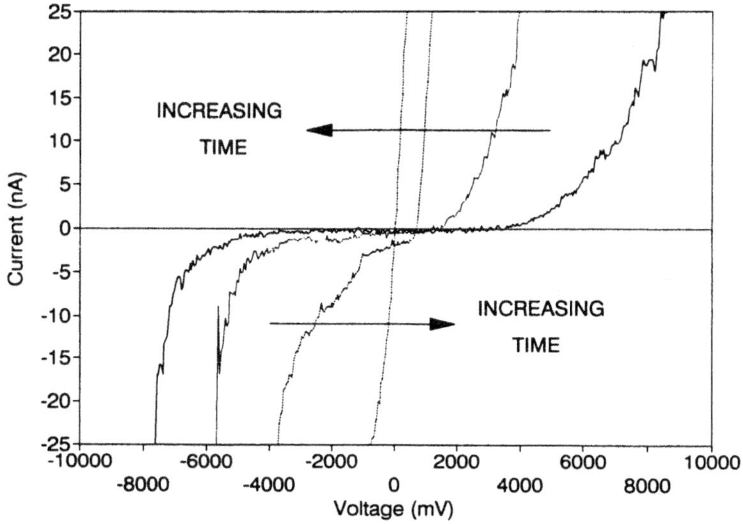

Figure 2. I/V curve changes as a function of time exposed to HCl. Note the narrowing passive region and increasing slopes of the regions with increasing time.

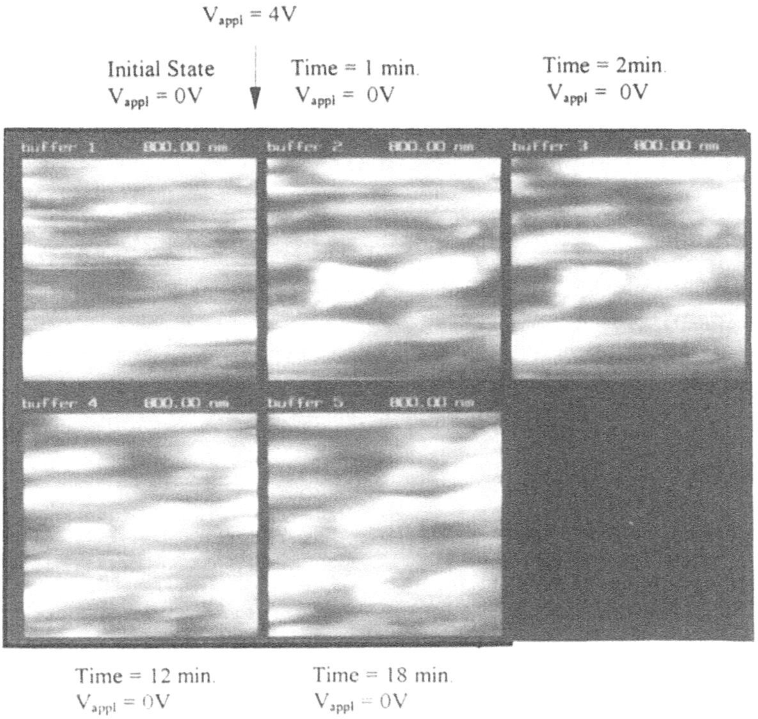

Figure 3. STM images, (a-e), show the passive film morphology changes as a function of applied voltage and time. (a) STM image of the initial state of the passive film; between (a) and (b) a potential of 4V was applied resulting in the centrally positioned film growth shown in (b), which was taken 1 minute after the applied potential was removed; (c),(d), and (e) STM images taken 2, 12, and 18 minutes, respectively, after the applied potential was removed.

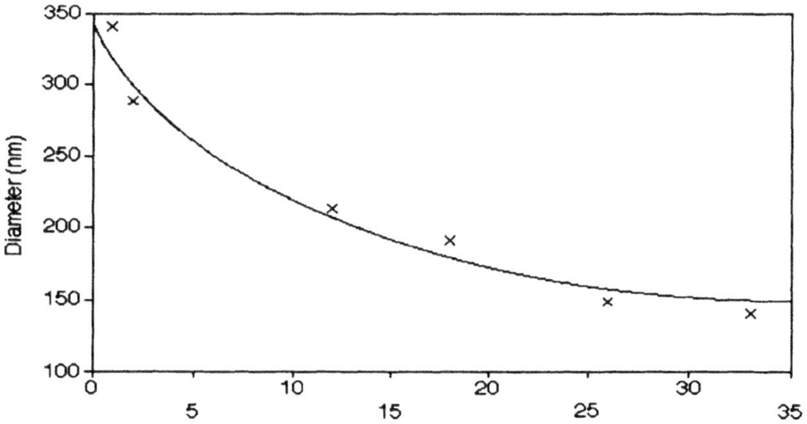

Figure 4. Diameter change of the oxide growth as a function of the time after the 4V potential was removed.

by measuring the diameter change of the growth with time. This curve shows that the decay of the growth is exponential in character. A nonuniform variation of the surface features, as observed, is expected where the passive film is controlled by charged entities, such as at dislocations and grain boundaries.

CONCLUSIONS

1. The STM can be used *in situ* to measure the extent of corrosion protection and stability of passive films exposed to etchants.
2. The STM can be used to establish quantitatively passive film changes at the nanoscale level, showing that the passive film is nonuniform.
3. A new model of passivation is given, which takes into account the structural properties of the oxide involved.

REFERENCES

1. M. Faraday, "Experimental Research in Electricity," vol. 2, London, England, 250-255 (1844).
2. G. C. Kiefer, "Surface Treatment of Metals," Edwards Brothers, Inc., Ann Arbor, MI, 1-25 (1941).
3. Mars G. Fontana, "Corrosion Engineering, 3rd ed.," McGraw-Hill, Inc., 469-481 (1986).
4. C.Y. Chao, L.F. Lin, and D.D. Macdonald, A Point Defect Model for Anodic Passive Films I. Film Growth Kinetics, *J. Electrochem. Soc.* 128: 1187-1193 (1981).
5. C.Y. Chao, L.F. Lin, and D.D. Macdonald, A Point Defect Model for Anodic Passive Films II. Chemical Breakdown and Pit Initiation, *J. Electrochem. Soc.* 128: 1194-1198 (1981).
6. C.L. McBee and J. Kruger, Localized Corrosion, R.W. Staehle, B. F. Brown, J. Kruger, and A. K. Agrawal, eds., NACE, Houston, TX, 252-258 (1974).
7. C.Y. Chao, L.F. Lin, and D.D. Macdonald, A Point Defect Model for Anodic Passive Films II, Chemical Breakdown and Pit Initiation, *J. Electrochem. Soc.* 128: 1194-1198 (1981).
8. R. Labusch and W. Schroter, Electrical Properties of Dislocations in Semiconductors, Dislocations in Solids, North-Holland Publishing Company, 127-191 (1983).
9. The Scanning-Probe Microscope Book, Burleigh Instruments, Inc., Fishers, NY, 1-18 (1990).

APPLICATION OF MAGNETIC FORCE MICROSCOPY IN MAGNETIC RECORDING

Thomas L. Altshuler

Advanced Materials Laboratory, Inc.
Concord, MA 01742-2112

Abstract: With the recent developments of scanning probe microscopes (SPMs), there has been a growth of knowledge about the surfaces of magnetic recording devices. The topography of various mass storage media has been examined in terms of roughness as well as identifying features of the surface. Magnetic track widths and bit widths have been observed and measured showing the differences in resolution as a function of lift height of the magnetic probe. The gap width, surface smoothness and other features of magnetic recording heads have been measured.

The exploding growth of the capabilities of scanning probe microscopes are making these instruments increasingly valuable in research, development, and manufacture of magnetic recording products.

INTRODUCTION

Magnetic recording devices have been used over the last 50 years for entertaining, replacing phonographs, and for other purposes, such as computer mass storage. This type of equipment consists of magnetic storage media, such as magnetic tapes or disks, and magnetic recording and reading devices, such as tape heads or computer heads. A typical head consists of a yoke of magnetic material with a gap, similar to a horseshoe magnet.

For an inductance-type head, there is an electric coil wrapped around the yoke, which will magnetize it when an electric current flows through the coil. A magnetic field will then pass between the two pole tips of the electromagnet across the gap. If the gap is close to the magnetic storage media, the magnetic field will magnetize a strip of material, called a "bit," in the media, which will retain its magnetization. The direction of magnetization of the "bit" depends upon the direction of the current flow in the coil, which controls the polarity of the field crossing the gap. When the gap is moved relative to the storage media while the current flow direction in the coil is changed, there will be "bits" of the media that are magnetized first in one direction and then in the opposite direction. In this manner, a series of "bits" can be stored, which makes up binary code information. These bits normally lie next to one another,

Atomic Force Microscopy/Scanning Tunneling Microscopy 2
Edited by S.H. Cohen and M.L. Lightbody, Plenum Press, New York, 1997

roughly perpendicular to the direction of travel of the head relative to the storage media, leaving a line of bits known as the "track."

Subsequently, this information can be read by passing the head over the "bits." For an inductance type head, a magnetic field magnetizes the head, which will in turn induce an electric current in the coil as the head traverses the bits. The magnitude of the current depends upon the speed of travel of the head over the bits, which influences the amount of magnetization of the head. The closer the head is to the storage media, the stronger the magnetization. Also, the smaller the bits, the more information can be stored per unit area of the storage media.

In order to improve the performance of magnetic storage devices, it is necessary to know details about the shape and magnitude of the magnetized bits in the storage media. Also, it is important to know the topography of the pole tips of the head, especially for a computer head, which does not contact a hard storage disk but flies close to the disk while the disk rotates at a high speed. The separation distance between the head and disk may be as small as a quarter of a micrometer.

With the invention of the scanning tunneling microscope by Binnig and Rohrer[1] and the subsequent invention of the atomic force microscope (AFM) by Binnig, Quate, and Gerber,[2] the surfaces of magnetic media and heads could be characterized in three dimensions. Altshuler,[1] used an AFM to measure the gap width of a computer read head. The subsequent development of the magnetic force microscope (MFM) and early experiments by Allenspach et al.[4] permitted examination of magnetic fields in various media using scanning tunneling microscopy with a flexible magnetized probe. This technique was also used by Hartmann et al.,[5] Martin and Wickramasinghe,[6] Rice and Moreland,[7] Murdock et al.,[8] and Gomez et al.[9] looked at high density recording with a flexible magnetic probe. Subsequent developments in magnetic force microscopy by Digital Instruments[10] have enabled the characterization of the magnetic field as a function of height above the media. One of their instruments has been used for this present work.

EXPERIMENTAL METHOD AND RESULTS

A Digital Instruments NanoScope™ III magnetic force microscope AFM was used in tapping mode to examine the surfaces of magnetic media and computer heads. Figure 1 shows the principle of operation in terms of the type of scans performed.

The top view of the scans shows that a scan is first performed with the magnetized AFM probe tapping on the surface of the specimen being examined. The scan first traces and then retraces in this contact mode. The side view of the scans shows the surface as observed in the retrace mode. The topography of the surface along that retrace scan is stored in the computer memory, which is subsequently used for the lift mode. The probe is then lifted a desired height above the surface and a second trace and retrace scan is made over the specimen surface. The mean position of the probe follows the topography of the surface that had been stored in computer memory. The probe is vibrated and also moves up or down in height until the frequency of vibration shifts from free resonance to a set frequency as a result of the magnetic field present. The greater the magnetic field, the greater the frequency shift for a given height. In this manner, magnetic domains can be seen.

Figure 2 shows an MFM scan of the surface of a standard VHS videotape. The left micrograph shows an image of the surface topography in which the lighter areas are higher than the darker ones. The right micrograph is the magnetic image of the same region. Here one can see the "bits" lying along vertical tracks.

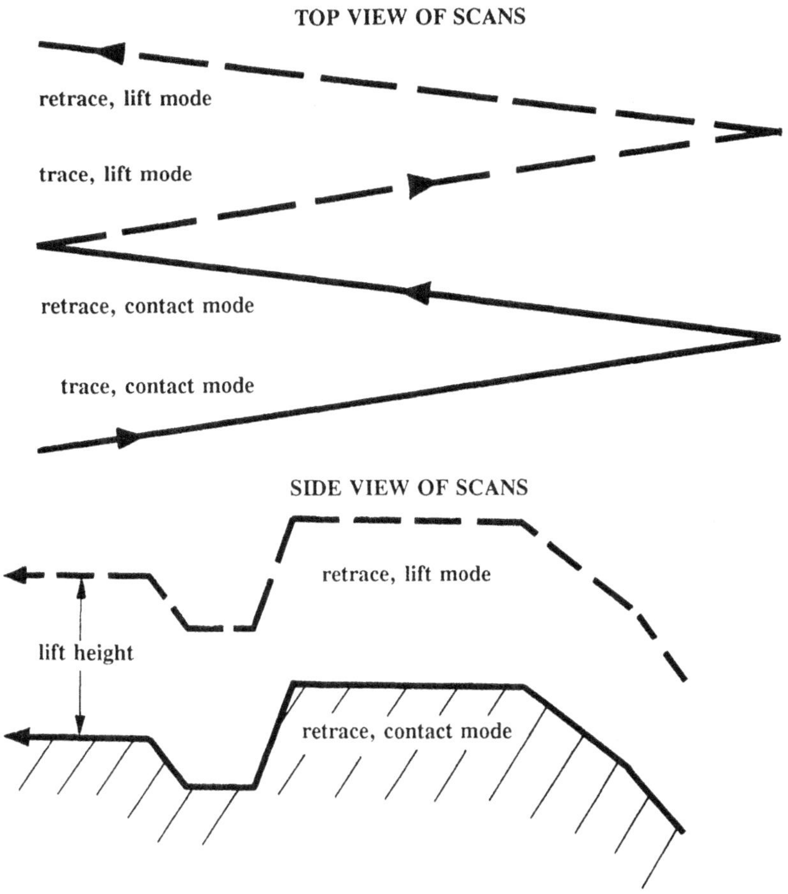

Figure 1. Scan process for magnetic force microscopy.

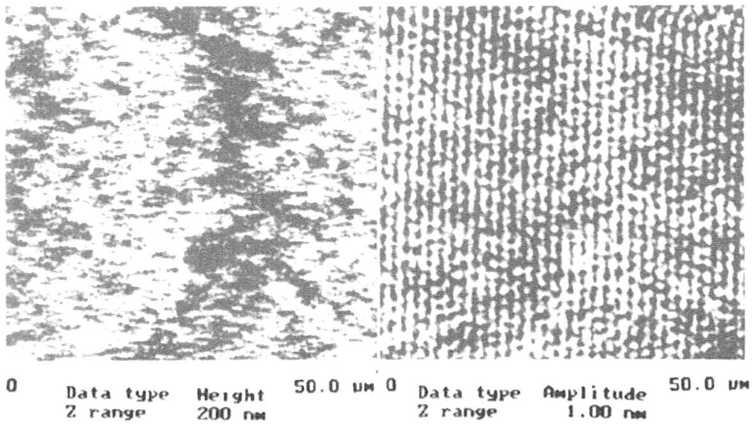

Figure 2. Surface and magnetic images of standard VHS videotape.

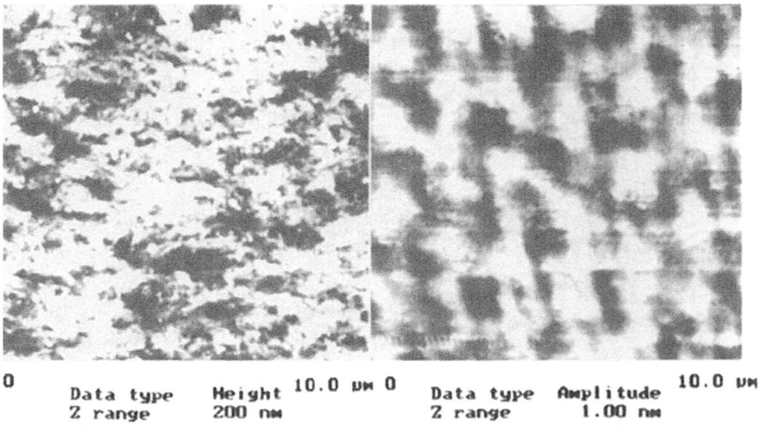

Figure 3. Magnified surface and magnetic images of standard VHS videotape.

By magnifying the center part of Figure 2, one can see the results in Figure 3. Here, the bits in the right micrograph are seen more clearly. There is no correlation between the surface features in the left micrograph and the magnetic image in the right micrograph, which proves that the magnetic image is not due to interference of the surface topography.

A magnetic image of a Sony ME High 8 videotape is shown in Figure 4. The surface is again seen in the left micrograph while the right micrograph presents the magnetic image of the bits and also the tracks. The tracks are the broad band of bits running from left to right. The bits are the long herringbone strips running approximately vertically.

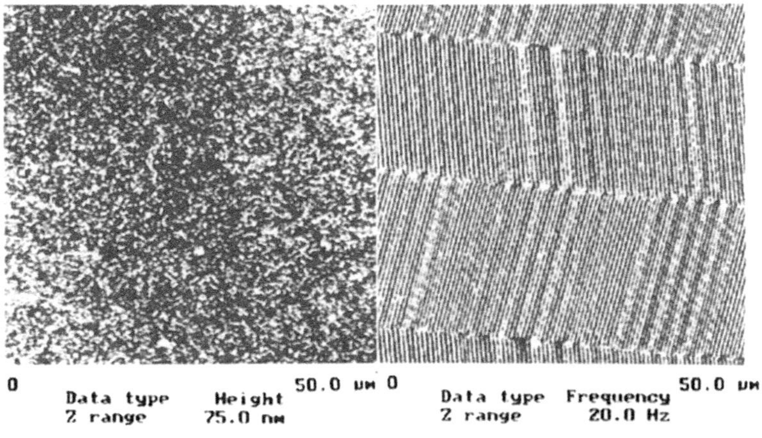

Figure 4. Surface and magnetic images of Sony ME Hi 8 videotape, bits and tracks.

The interface between the tracks when magnified can be seen in Figure 5. Here the bits are shown as the vertical light regions in the right micrograph with some disorder of the domains between the tracks. The left micrograph shows the surface with spherical lubrication nodules, which are the small light round objects.

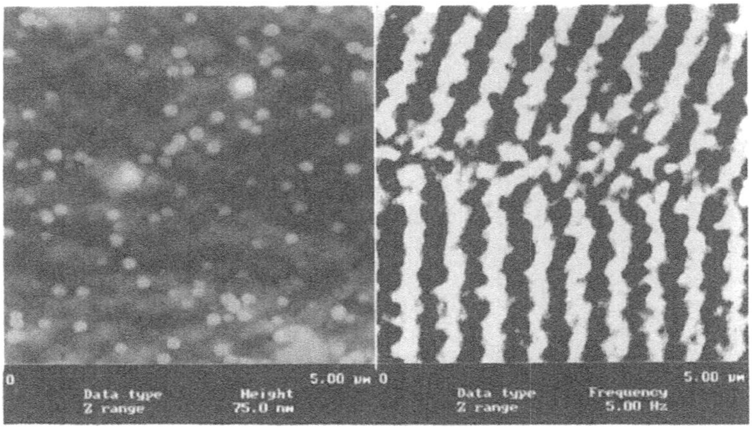

Figure 5. Surface and magnetic images of Sony ME Hi 8 video tape showing bits.

Prior to the development of the magnetic force microscope, a standard method of observing magnetic domains was with the use of ferrofluid. Figure 6 shows ferrofluid placed on a Digital Equipment Corp. (DEC) computer hard disk. This disk, magnetically written by George Islay[11] and provided by Philip Bartels[12] was an Head Stack Assembly (HSA) SHR 23 disk written by Proto Drive # EL 3440025 at a frequency of 23.73 MHZ with a rotation speed of 7200 rpm at a radius of 1.0553 inches. The write current was 20 mA. The speed was 8.33 milliseconds/bit. The tracks can be seen in the micrograph, but the bits are not discernable. The tracks seen in Figure 6 were then observed with an MFM (see Figure 7). There was a Knoop hardness indentation placed near the tracks for location purposes. The left micrograph

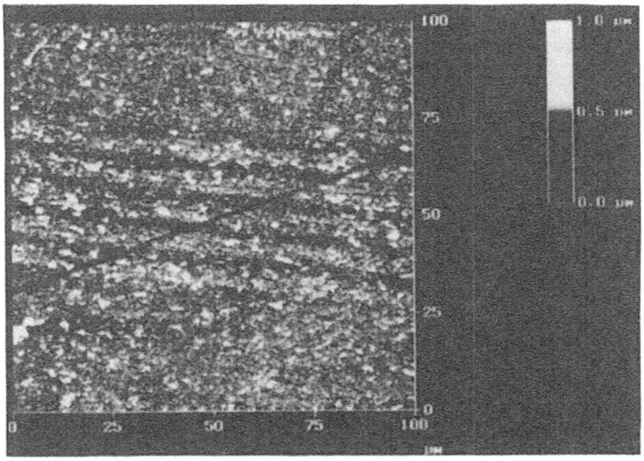

Figure 6. AFM top view of DEC hard disk showing recorded tracks with ferrofluid.

presents the surface topography of the disk where the surface was indented (dark center region) with light regions on either side of the indentation where the metal was plastically deformed above the surrounding disk surface. The right micrograph shows the tracks quite clearly as well as the individual bits.

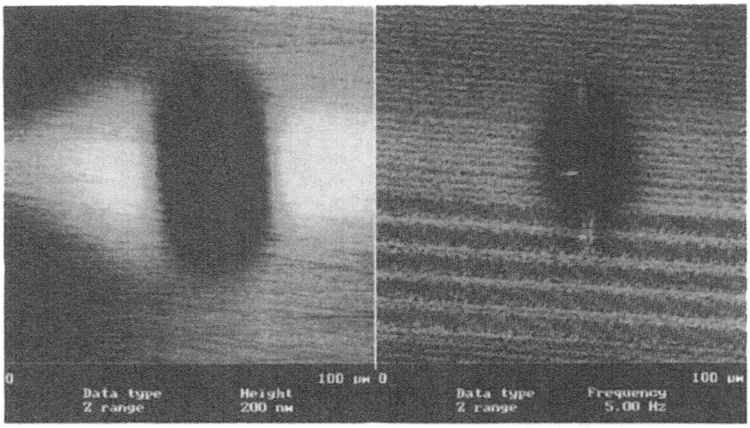

Figure 7. Surface and magnetic images of DEC hard disk showing bits and tracks.

Upon magnification, Figure 8 shows the surface of the disk in the left micrograph. the scratches on the surface prevent the head from wringing onto the disk surface when in contact with that surface. The right micrograph shows a track running from left to right with the individual bits aligned vertically. The lift height of 200 nanometers (nm) resulted in a weak magnetic field where the bit edges are not defined clearly. When the lift height is lowered to 100 nanometers, the bits are defined much more clearly, as can be seen in Figure 9. The cause of the curved ends of the bits, giving each bit an "S" shape, is not understood clearly at this time.

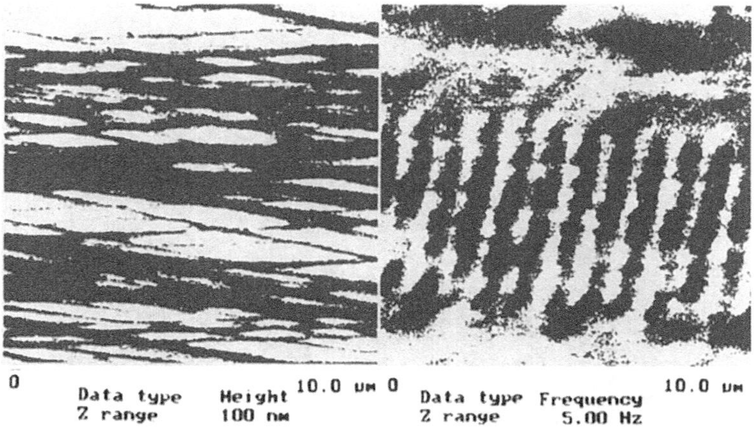

Figure 8. Surface and magnetic images of DEC hard disk, lift height 200 nm.

A cross-sectional view of the magnetic image of Figure 9 can be seen in Figure 10. Here, the peak to valley distance of the magnetic field can be measured, although a quantitative correlation between these values and the actual field has not been determined. The cross-section line is shown in white across the micrograph at the lower left hand corner of the Figure. Arrow pairs are placed at the peak and valley of bits in the magnetic image with the dimensions shown in the table at the lower right-hand corner of the figure.

Lowering the lift height further to 50 nanometers results in Figure 11. Although the bits do not stand out as units as well as in Figure 10, details of each bit are shown with greater resolution. Also the nonmagnetized part of the disk with the patterns of the individual domains are revealed quite clearly. From these examples, one could obtain a three- dimensional picture of the magnetic field above the disk surface.

Figure 12 shows a Ferroxcube™ read head used for computer tapes. A Knoop indentation was placed across the glass gap for reference purposes. The individual grains of MnZn hot pressed ferrite can be seen in which the texture changes with the crystallographic orientation of the grains as a result of polishing. The surface roughness for the entire surface viewed, as well as that within the white box in the upper left of the micrograph, is presented in the table.

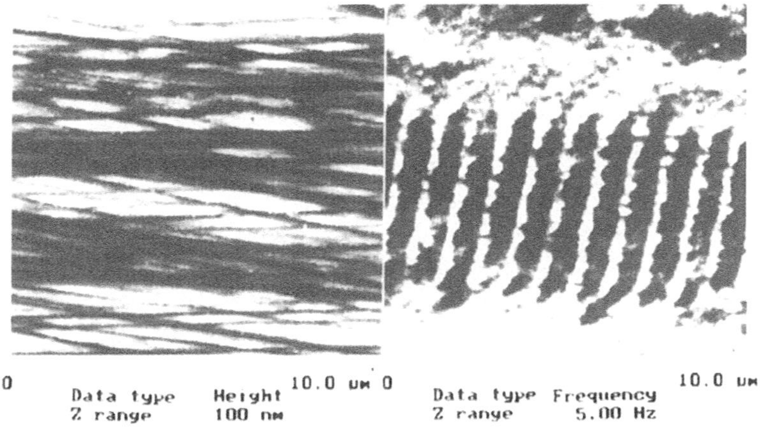

Figure 9. Surface and magnetic images of DEC hard disk, lift height 100 nm.

By magnifying the glass gap of the head, the shape of the gap and its dimensions can be measured, see Figure 13. Also the texturing of the grains on either side of the gap can be observed.

A computer thin film head was examined with the AFM as shown in Figure 14. The two pole tips can be seen as the rectangular objects in the lower part of the micrograph. A Knoop indentation was made for reference purposes. Note that the left pole tip is recessed below the surrounding alumina layer while the right pole tip is about the same height as the surrounding alumina. This recession would degrade the magnetic field slightly when writing onto a disk.

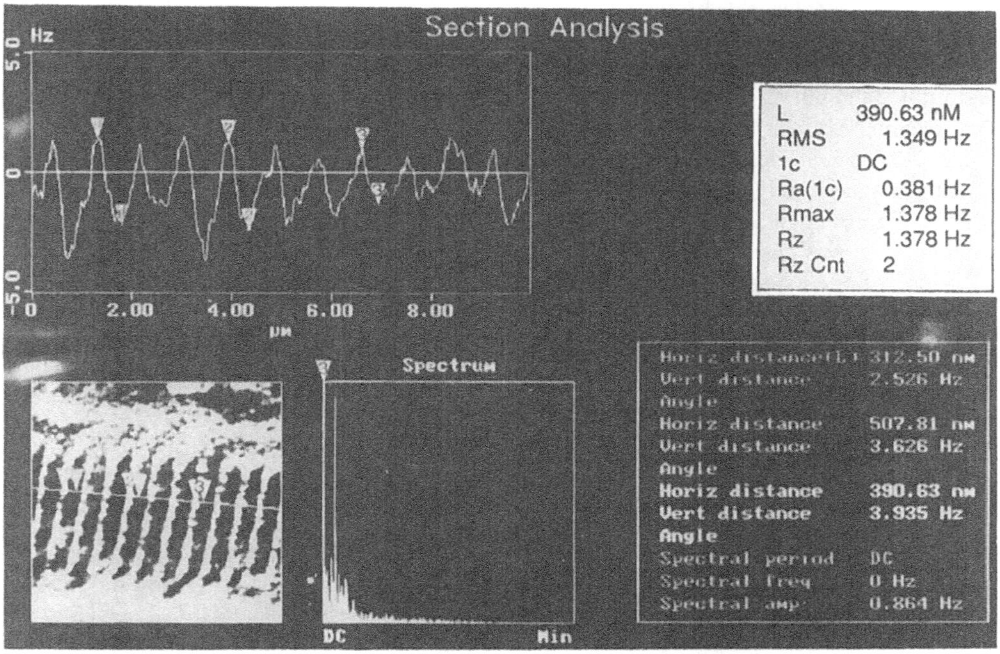

Figure 10. Cross section of magnetic images of DEC hard disk, lift height 100 nm.

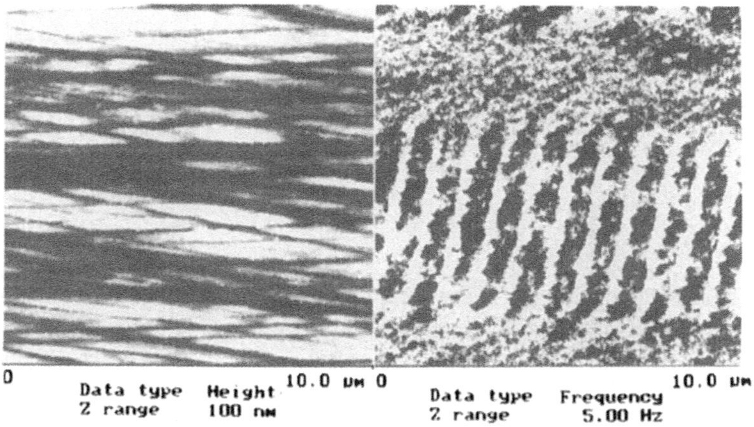

Figure 11. Surface and magnetic images of DEC hard disk, lift height 50 nm.

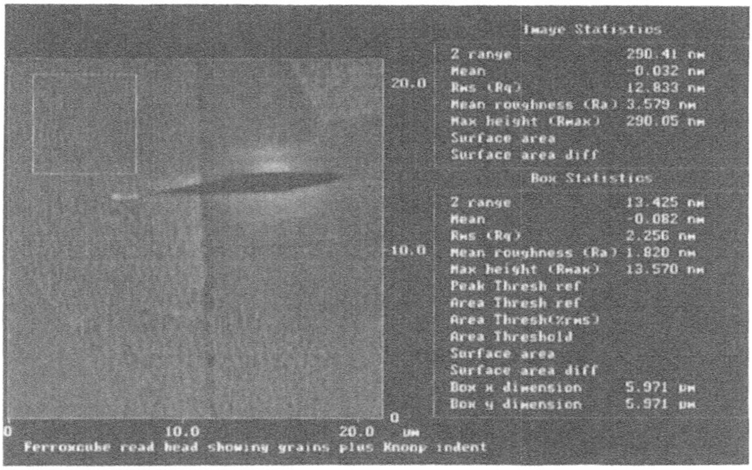

Figure 12. Top AFM view of Ferroxcube read head.[3] Roughness analysis.

To determine the amount of pole tip recession shown in Figure 14, a cross-sectional view was made; see Figure 15. The arrows show that the left pole tip was recessed 6.2 nanometers. An example of pole tip protrusion of a computer thin film head can be seen in Figure 16. Here, the right pole tip is protruding above the surrounding alumina. The Knoop indentation is used for reference.

The pole tips in Figure 16 were viewed in a cross-sectional presentation in Figure 17. The right pole tip in this case is protruding by 13.8 nanometers above the surrounding alumina. Also, the alumina was lapped 11.7 nanometers below the Alsimag ceramic to the left of the micrograph.

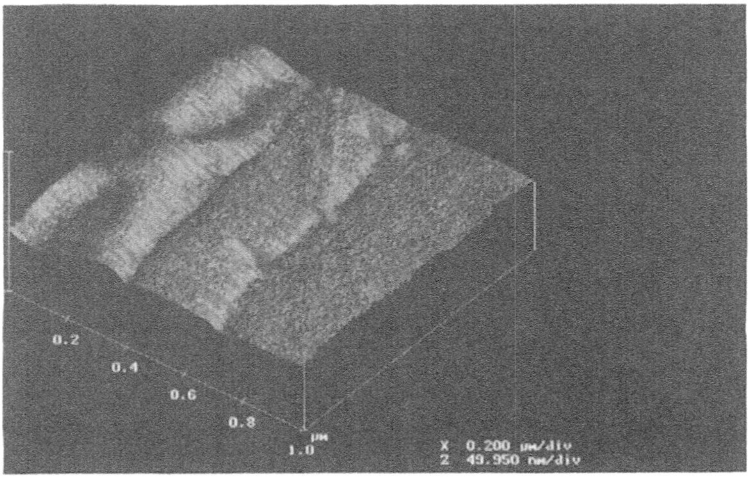

Figure 13. Surface AFM view of Ferroxcube read head showing glass gap 280 nm.[3]

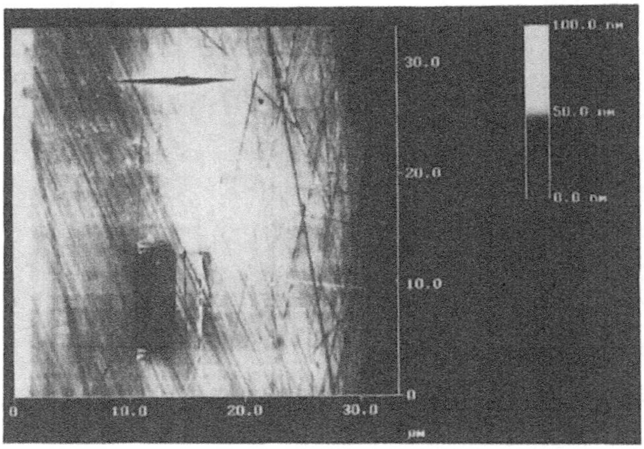

Figure 14. Top AFM view of thin film head showing pole tip recession.

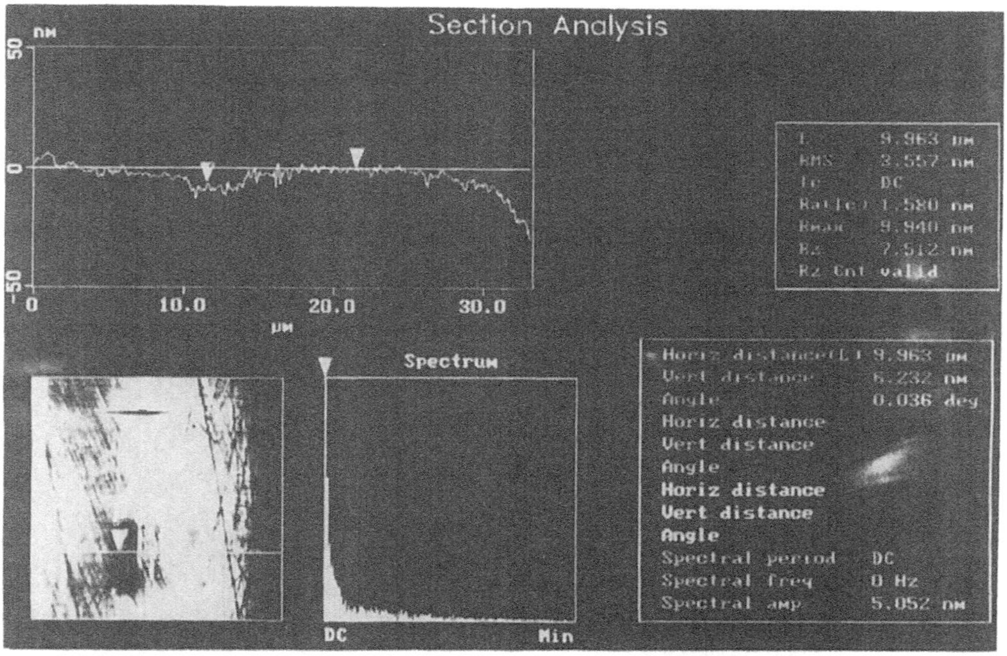

Figure 15. Cross section AFM view of thin film head showing pole tip recession.

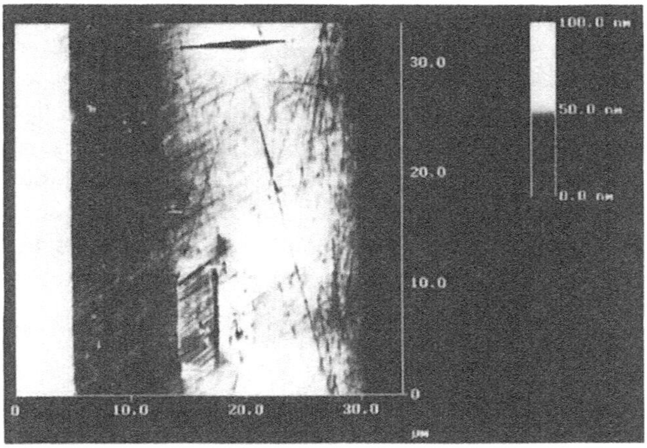

Figure 16. Top AFM view of thin film head showing pole tip protrusion.

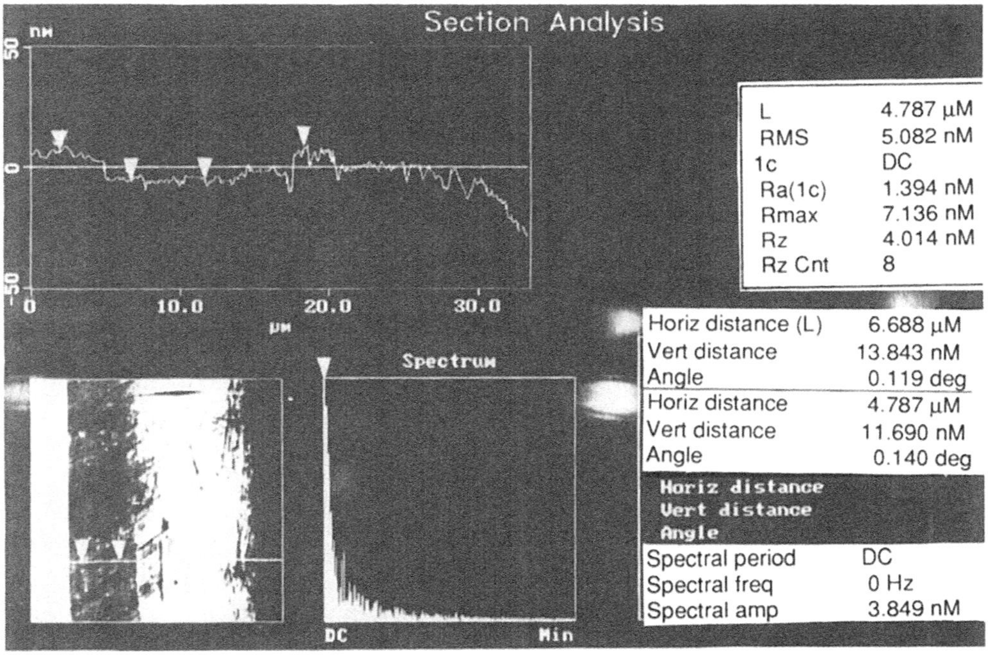

Figure 17. Cross-section AFM view of thin film head showing pole tip protrusion.

CONCLUSIONS

Developments in the techniques of magnetic force microscopy from scanning probe microscopes have enabled examination of the magnetic fields associated with magnetic recording devices. Magnetic images have been obtained of the written information in magnetic tapes and computer hard disks. Resolution has been improved considerably when observing "bits" and "tracks" by the magnetic force microscope MFM compared to those features revealed by ferrofluids. In addition, changes in the magnetic images obtained from different lift heights of the MFM provided three-dimensional information of the magnetic field, which is not possible with other techniques, such as with the use of ferrofluids or Kerr rotators.

The atomic force microscope has been used to study the surface topography of both storage media and recording heads. This is important for developing better manufacturing techniques of these components for improved magnetic performance.

REFERENCES

1. G. Binnig, H. Rohrer, C. Gerber, and E. Weibel, Surface study by scanning tunneling microscopy, *Phys. Rev. Lett.*, 49: 57-60 (1982).
2. G. Binnig, C.F. Quate, and C. Gerber, Atomic force microscope, *Phys. Rev. Lett.*, 56: 930-933 (1986).
3. T.L. Altshuler, Atomic-scale materials characterization, *Adv. Mater. Process.*, 130, No. 3: 18-23 (1991).
4. R. Allenspach, H. Salemink, M. Bischof, and E. Weibel, Tunneling experiments involving magnetic tip and magnetic sample, *Z. Phys.* B. 67:125-128 (1987).
5. U. Hartmann, T. Göddenhenrich, H. Lemke, and C. Heiden, Domain-wall imaging by magnetic force microscopy, *IEEE Trans. Magn.* 26: 1512-1514 (1990).
6. Y. Martin and H.K. Wickramasinghe, Magnetic imaging by "force microscopy" with 1000 Å resolution, *Appl. Phys. Lett.* 50, 20: 1455-1457 (1987).
7. P. Rice and J. Moreland, Tunneling-stabilized magnetic force microscopy of bit tracks on a hard disk, " IEEE *Trans. Magn.* 27: 3452-3454 (1991).
8. E. Murdock, R. Simmons, and R. Davidson, Roadmap for 10 gbit/in^2 recording media: challenges, IEEE *Trans. Magn.* 28: 3078-3083 (1992).
9. R. Gomez, A. Adly, I. Mayergoyz, and E. Burke, Magnetic force scanning tunneling microscopy: theory and experiment, *IEEE Trans. Magn.*, 29: 2494-2499 (1993).
10. Digital Instruments Inc. (manufacturer): NanoScope™ III Stand Alone MultiMode™ Atomic Force Microscope.
11. George Islay, Digital Equipment Corporation, 333 South Street, Shrewsbury, MA 01545.
12. Philip Bartels, Digital Equipment Corporation, 333 South Street, Shrewsbury, MA 01545.

SCANNING ELECTRON MICROSCOPY, SCANNING TUNNELING MICROSCOPY, AND ATOMIC FORCE MICROSCOPY STUDIES OF SELECTED VIDEOTAPES

Ernest C. Hammond, Jr.

Morgan State University
Department of Physics
Baltimore, MD 21239

Abstract: This study hopes to examine the various classification protocols as well as manufacturer brand videotapes, to delineate selected characteristic physical changes within the metallic oxide on the tapes at the nanometer regime to include the domain levels. Moreover, an examination of the tape magnetic domain boundaries and other particle characteristics from tape to tape and manufacturer to manufacturer will utilize the scanning tunneling microscope (STM), scanning electron microscope (SEM), and the atomic force microscope (AFM). This study examines these traits and characteristics by comparing and contrasting the results obtained from the STM, the AFM and the SEM for nonrecorded tapes.

INTRODUCTION

Since the advent of audio and videotapes, the tape classification system as well as price differentials associated with these classifications have existed.[1] Classical experiments have shown that constant frequencies recorded on magnetic tapes have dominant trait changes visible with the SEM and AFM.[2] The use of videotapes in our daily lives is well documented.[3] The real question concerning the videotape use is which tapes are better to buy and use? The critical issue is whether the videotape grading system has a physical effect on the performance of the tape. Television professional cameramen have observed that there are more drop-out effects observed when the cheaper grades of tapes are used.[4] This research team has initiated an examination of two grades of videotape. They include the Polaroid Super color T-120 and the Sony high grade videotapes. The important feature of both tapes is that they are both new tapes with no electrical signals on them. The tapes were subjected to SEM, STM, and AFM analysis.

Atomic Force Microscopy/Scanning Tunneling Microscopy 2
Edited by S.H. Cohen and M.L. Lightbody, Plenum Press, New York, 1997

MATERIALS AND METHODS

The videotapes used were the Polaroid Super Color T-120 and the Sony High Grade videotapes. The tapes were not coated or processed in any manner. The tapes were fresh from the sealed box and opened to get a piece of tape of appropriate size for analysis using the scanning electron microscope initially, followed by images obtained from the atomic force microscope and the scanning tunneling microscope. The initial phase of our examination was to examine the tape without magnetic imprints.

RESULTS

The SEM analysis indicated a variation of grain structure with the appearance of hole like cratering on the tape. Moreover, the physical observation demonstrates the tape is highly magnetic in nature, sticking to the pole pieces of the STM. An x-ray analysis using energy dispersive spectroscopy (EDS) demonstrates (Figure 1) the tape contains iron and chromium compounds. We have observed large white grains indicating very high conductivity of these grains from secondary and back scattered electrons (Figures 2-5). The Polaroid cheaper grades (Figures 6-8) have elongated grain structures, while the Sony High Grade has more circular grain structures.

An attempt was made to use the STM on the samples of tape using both highly ordered pyrolytic graphite (HOPG) as a source of tunneling current and direct contact with the IR/Pt tip near contact with that videotape; neither case indicated any tunneling effect.

Using the STM is a challenging activity and this researcher wishes to revisit this unique problem. It should be noted that at very high magnification the white grains are conducting and producing large numbers of back scattered and secondary electrons, which are visible on the (SEM Figures 2-4, 6-8). The anisotropy of the Polaroid T-120 grains is clearly visible, when compared with the Sony High Grain tape.

The AFM produces very satisfactory views of the grains and quantitative capability to measure grain heights; the anisotropy is more vividly seen when using Polaroid T-120 tape (Figures 9, 10). Moreover, both tapes show the cross-sectional surface heights using the Digital III AFM. At a higher power, one sees the individual grains protruding continue to display the elongated grain effects associated with the Polaroid T-120 (Figures 11, 12, 13). The variations in height within crevices vary from 2000 Å for the Sony High Grade to the 1500 Å for the Polaroid T-120. The height differences between the Sony High Grade and the Polaroid T-120 is an average 500 Å.

Many of the tape companies are doing proprietary research activities on grain structure analysis using the atomic force microscope on various video and audio tapes.[5]

Examination of audio and video grains at the nano-regime may have an effect on the reduction of aging effects within these tapes caused by thermal and other external fields. Moreover, the high-grade tape having deeper crevices than the T-120 Tape may explain the minimal drop-out phenomena experienced by professional videographs.

CONCLUSION

Figures 2 and 3, 1.4x magnification and 3.9 magnification respectively, represent the same area of a piece of Sony High Grade videotape.

It is noticed that with the scanning electron figures the valleys between the grains and the grains themselves display limited anisotropy. Figure 4 at 29,000x magnification will show

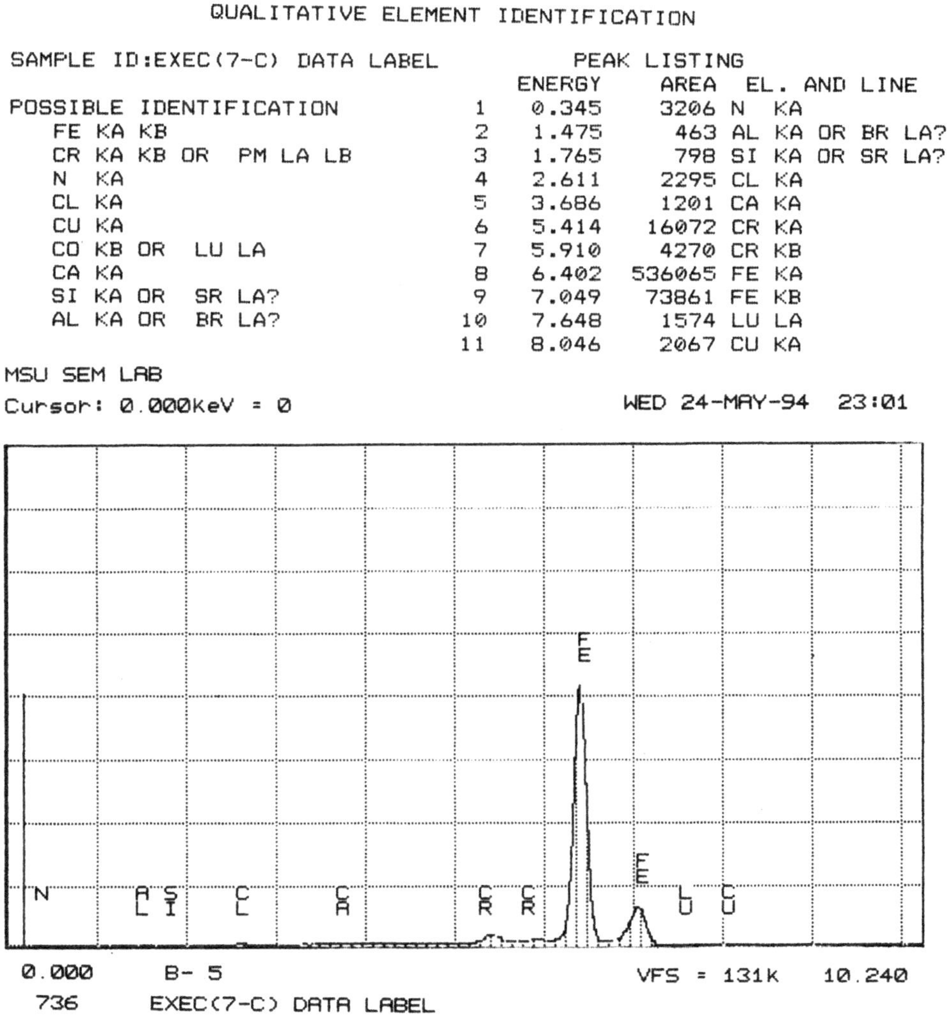

Figure 1. Qualitative element identification.

that the applications of the grains to the acetate backing seem to have limited anisotropy. Now an examination of another section of the tape using an atomic force microscope in both the 2-D and 3-D mode substantially confirms the view that the grains have limited anisotropy (Figures 9, 10) while in a 3-D projection one can see that the grain tops are relatively circular or elongated with no particular preferred direction in terms of the height. Typically, we were looking at average grain heights almost 1700 Å high, which are magnetized while the tape is running.

The real question is, "How much of these 1700 angstroms average height contributes to the wear and tear of the magnetic head due to the frictional motion associated with the taping or playback process?

Using STM, the T-120 videotape seems to display an elongated grain anisotropy, even at 20,000x magnification (Figure 7). The grains and crevices within the tape are quite elliptical or oblong in shape (Figures 6,8,14). Even using the atomic force microscope over the selected area, which has no electronic signal on it, the grains are profoundly longer and even a brief view seems to indicate that there is an associated double peak as viewed with AFM (Figures 11,12). By using the same field of view of the SEM, and looking at the surface of the Polaroid T-120 tape, there are clearly preferential directions associated with the grain structures. This anisotropy of the grains is even more pronounced at 4000x using an SEM (Figures 7, 8, 14). A zoom image of this T-120 surface indicated bubbly boundaries and valleys that seem to give rise to this anisotropy within the domain formations (Figure 13). This may be the reason why many expert cinematographers and videographers are quite sensitive to the drop-out phenomenon[4] that is associated with the use of cheaper grades of videotapes.

Using EDS for x-ray analysis of the grains on the tape, there are peaks of chromium, cobalt, and copper (Figure 1) and the aluminum and silicon peaks are associated with the sample stub and the silicon binder used to hold the sample tape in place.

An emission spectra of the Sony High Grade videotape and the Polaroid T-120 videotape from 250 Å through 800 Å (Figure 15(a)) indicates that the Sony High Grade tape has higher intensity of spectral lines with as many as 20 peaks through out the visible region, while Polaroid is missing structure and intensity.

Using an excitation beam of 300 nm, the comparative excitation spectra continues to demonstrate that the High Grade tape has more intense peaks from 200 nm to 400 nm (Figure 15(b)) indicating more electronic responsiveness than the T-120 Polaroid. Major peaks of the Sony and Polaroid Tapes do mirror themselves throughout both spectra.

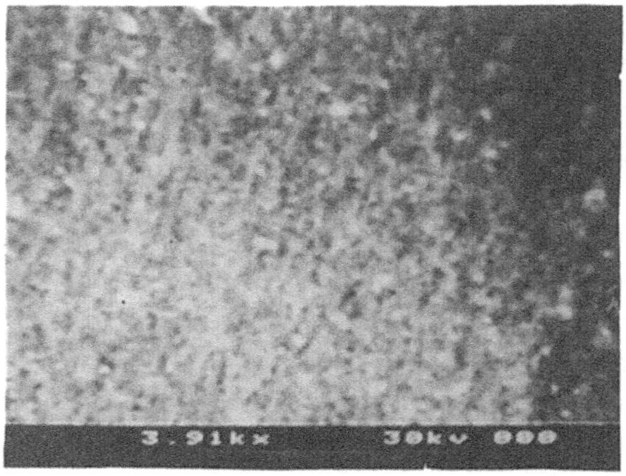

Figure 2. SEM micrograph 1 Sony High Grade 3.9kX.

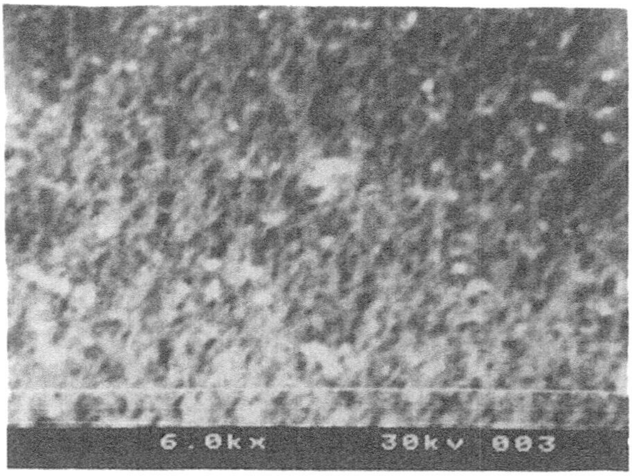

Figure 3. SEM micrograph 2 Sony High Grade 6.0kX

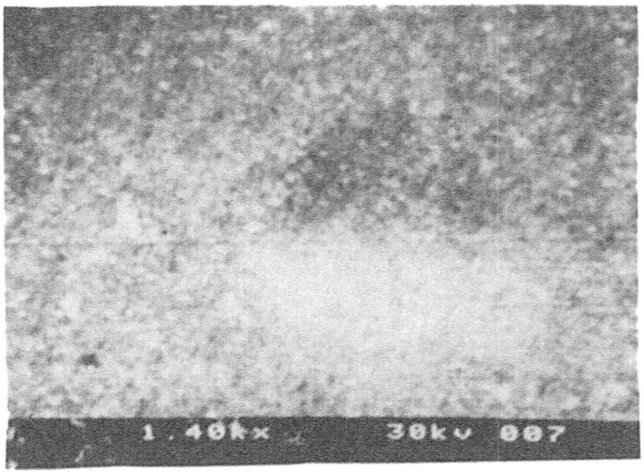

Figure 4. SEM micrograph 3 Sony High Grade 1.4kX.

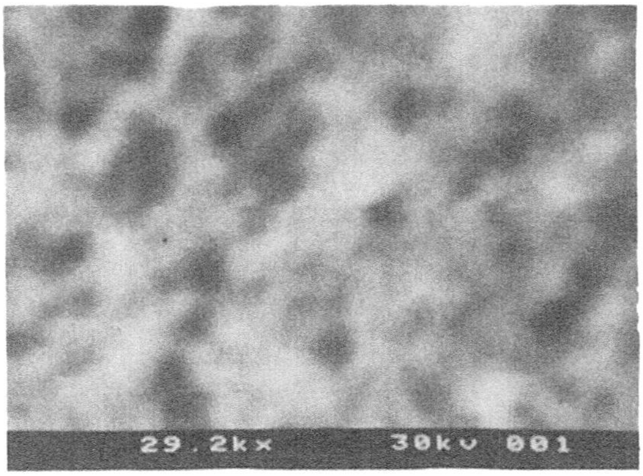

Figure 5. SEM micrograph 4 Sony High Grade 29.2kX

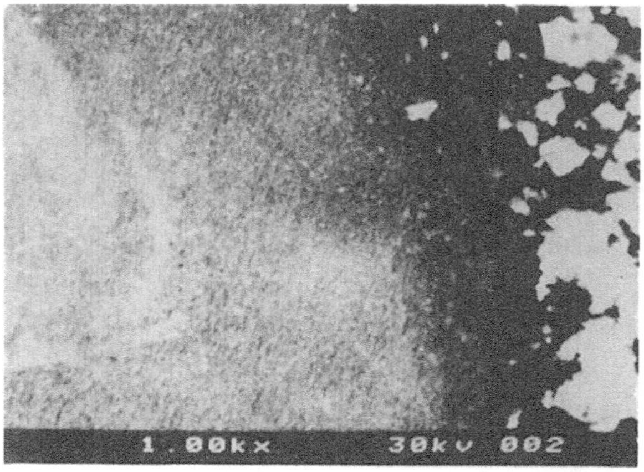

Figure 6. SEM micrograph 7 Polaroid T-120 I.0kX

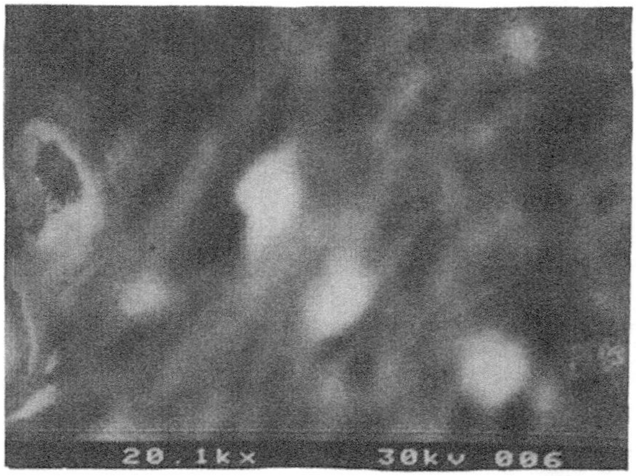

Figure 7. SEM micrograph 8 Polaroid T-120 20.1kX

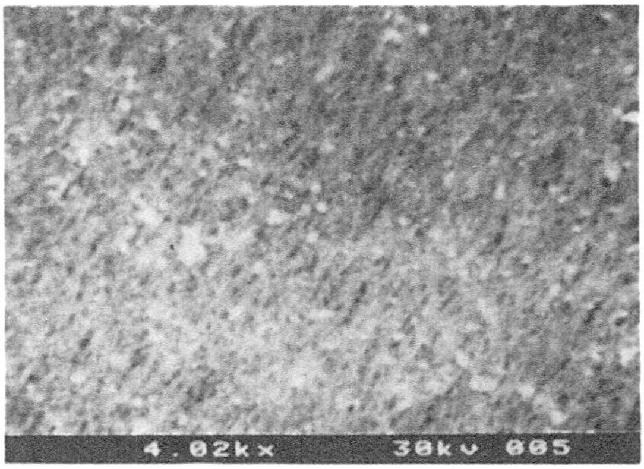

Figure 8. SEM micrograph 9 Polaroid T-120 4.02kX

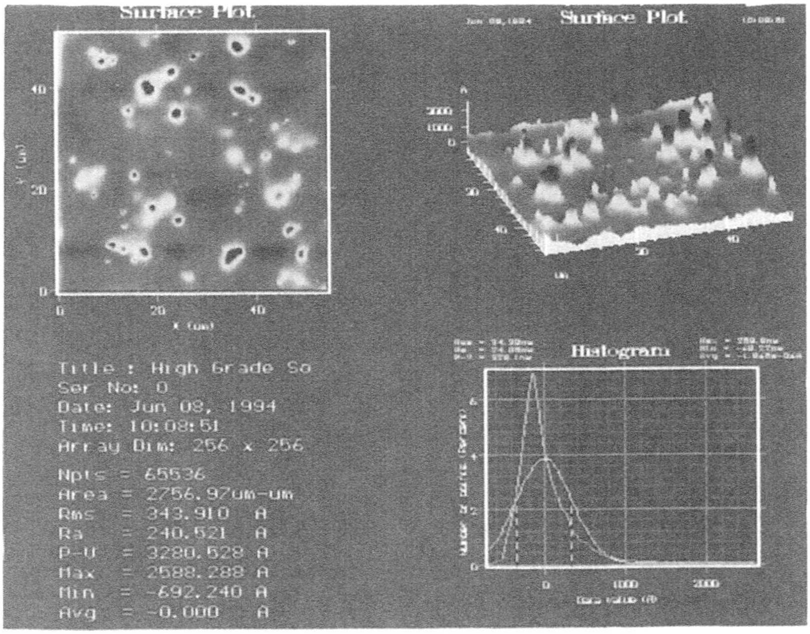

Figure 9. AFM micrograph 5 Sony High Grade

Figure 10. AFM micrograph 6 Sony high grade.

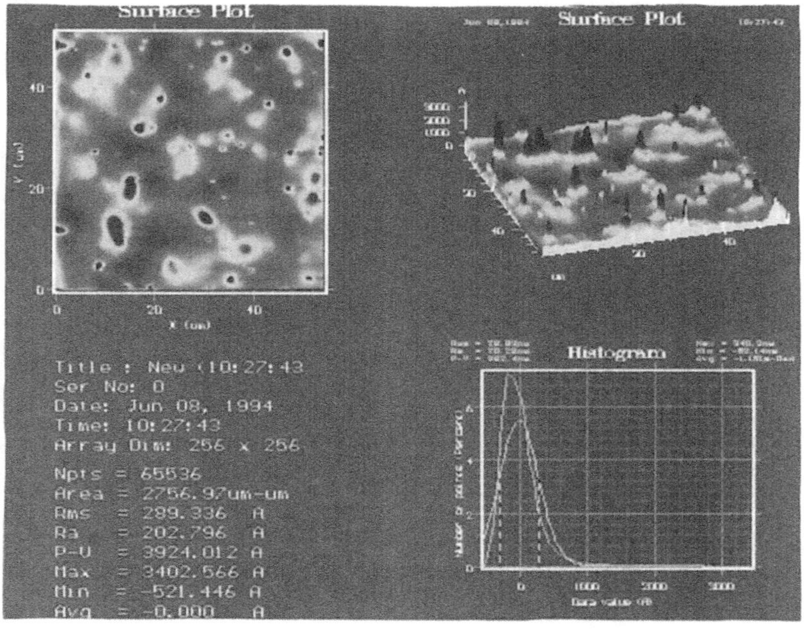

Figure 11. AFM micrograph 11 Polaroid T-120.

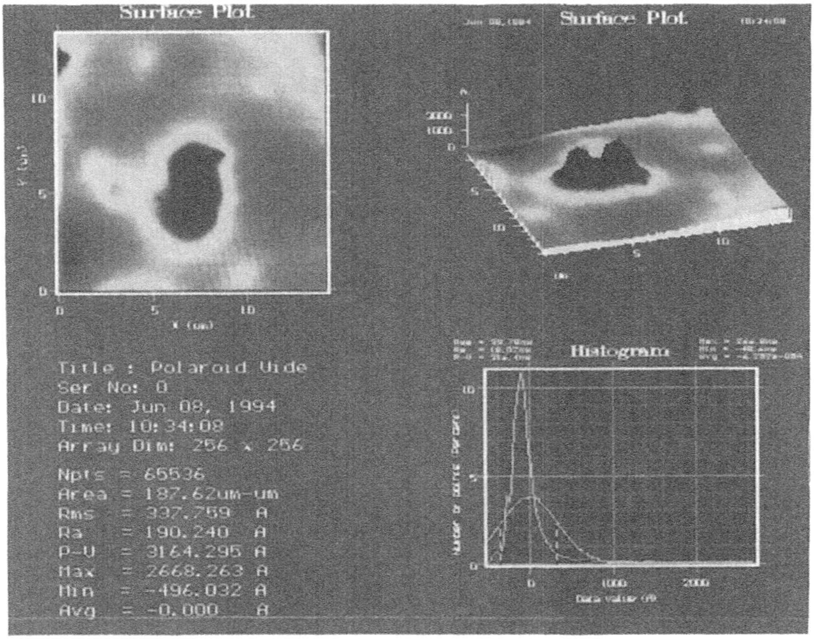

Figure 12. AFM micrograph 12 Polaroid T-120.

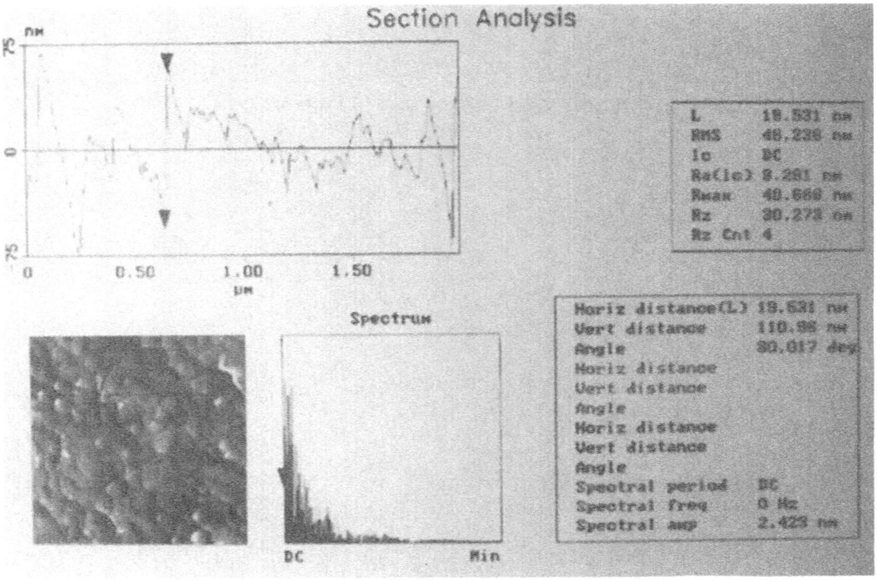

Figure 13. AFM micrograph 13 Polaroid T-120.

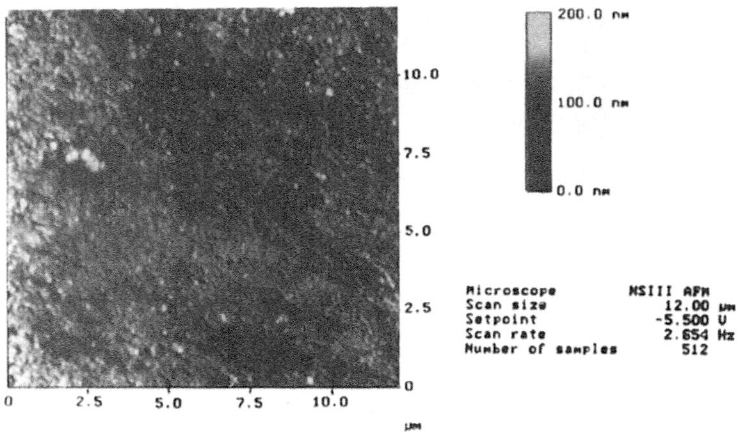

Figure 14. AFM micrograph 10 Polaroid T-120.

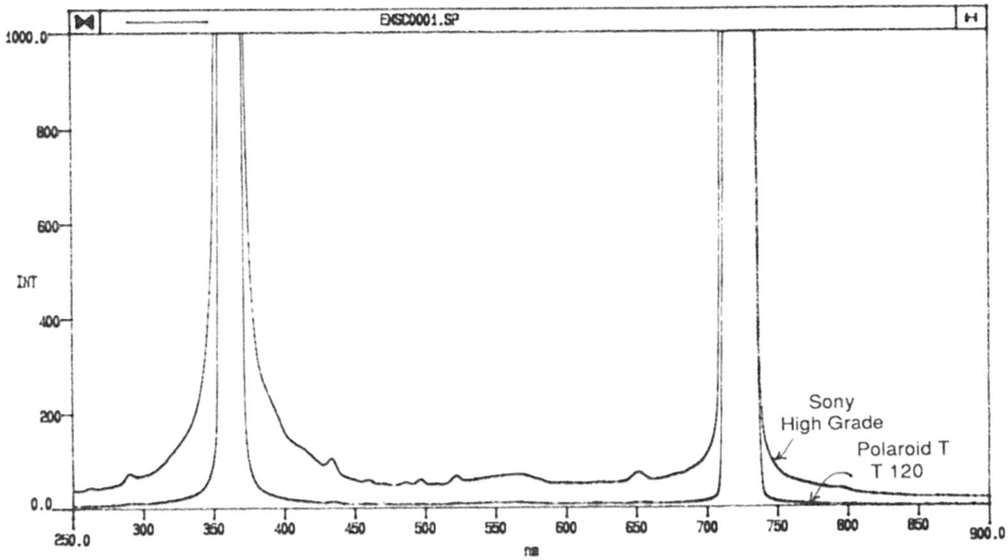

-Filename	#pts	start	end	min	max	res	ord acc
EMSC0001.SP	1301	250.00	900.00	18	1000	0.16	INT 1

262.0	44	266.5	42	290.0	74	295.0	67	351.5	1000
427.5	91	433.5	104	454.5	56	459.0	60	479.0	48
485.5	54	490.5	51	496.0	60	503.0	50	506.5	51
511.0	49	522.0	68	529.0	58	564.5	71	591.0	52
596.5	53	610.0	51	622.5	54	629.5	53	651.0	74
661.0	59	710.0	1000						

Figure 15. (a) and (b). The emission spectra of the Sony High Grade Videotape and the Polaroid T120 less expensive tape were compared. At certain frequencies the Sony High Grade tape had very definite peaks and higher intensity readings. The two tall peaks around 51.1 nm and 710 nm represent the excitation frequencies of the quartz windows.

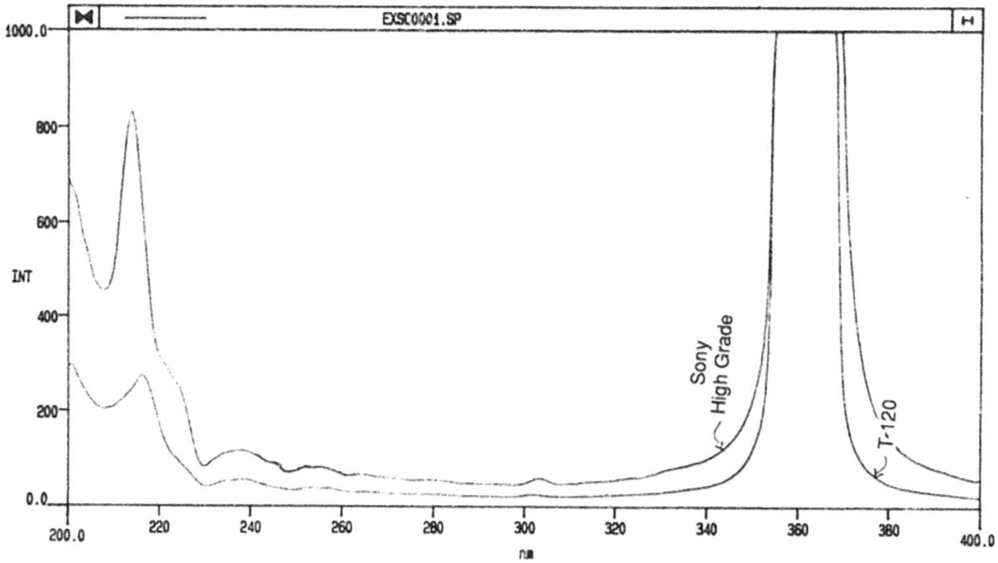

-Filename	#pts	start	end	min	max	res	ord	acc
EXSC0001.SP	401	200.00	400.00	22	1000	0.16	INT	1

200.5	298	208.0	203	216.0	275	230.5	42	238.5	57
249.0	35	253.0	40	262.5	31	265.0	32	296.5	22
301.5	27	308.0	22	355.0	1000				

Figure 15. (b). See Figure 15 (a).

REFERENCES

1. Classification system is shown on the back of most VHS tapes for Sony as well as the Polaroid VHS tapes.
2. Robert Cassidy, Magnetic force "attracts" users, The Third Annual Genius Issue, *Research and Development Magazine*, 18-19, Nov. 1995.
3. Lawrence M. Fisher, Technology memories linger, but the tape fades, *The New York Times*, 9, Nov. 28, 1993.
4. Local TV cameramen have testified to the fact that in doing video work, there is more drop out occurring with cheaper VHS tapes.
5. Members of the major tape manufacturers have cited that major contracts have been let to do this kind of work and have stated that the details of this research are proprietary.

SURFACE CHARACTERISTICS EVALUATION OF THIN FILMS BY ATOMIC FORCE MICROSCOPY

G. Li

University of Massachusetts at Lowell
Center for Advanced Materials
Lowell, MA 01854

Abstract: The surface morphology of a variety of thin films such as diamond thin films and coated powders have been evaluated by atomic force microscope (AFM). This study demonstrates the use of AFM as a technique for the optimization of thin film deposition and coating process. A discussion for surface roughness evaluation is presented. Also, the force acting for roughness evaluation is discussed.

INTRODUCTION

AFM is capable of analyzing surface structure of materials, both conductors and insulators. It is especially useful for characterizing the surface of thin films in 2-D or 3-D morphology and to optimize coating process.[1-7] We have studied a variety of thin films. They include an incomplete diamond thin film and ZnS coated powder. The surface of all these materials is very rough. Their surface characterization results are as follows.

Incomplete Diamond Thin Films

The surface morphology of a diamond thin film is as shown in Figures 1 and 2. The polycrystalline diamond is a chemical vapor deposition (CVD)-deposited layer on a silicon substrate with incomplete coverage. Grain structures on a silicon substrate in the incomplete diamond sample are shown in Figure 1. A column as structure of the same sample is shown in Figure 2. The three parallel columnar structure is stretched to the small, incomplete diamond crystal layer. It is different from the grain structure reported in most papers regarding crystalline diamond growth process. The grain structure of diamond thin film nuclei is Δ tetrahedral with (111) bases and (100) sloping faces. Thus, columnar structure may be caused from a different mechanism of diamond thin film growth.

Because the surface of a diamond thin film is extremely hard and rough, the tip of the AFM is easily worn out during the scanning process. In addition, at different locations the

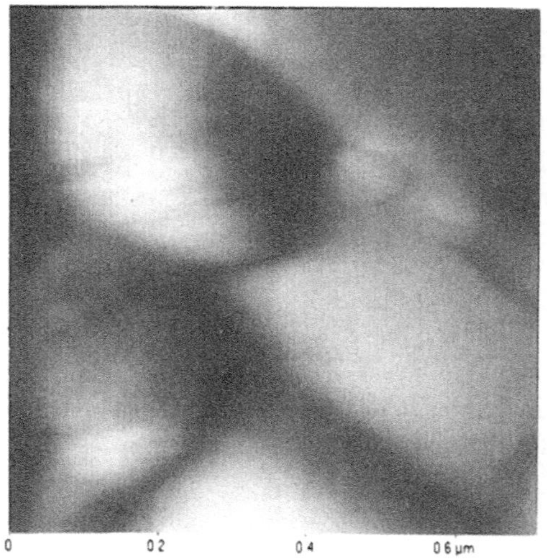

Figure 1. Granular structures of diamond thin films.

Figure 2. Columnar structures of diamond thin films.

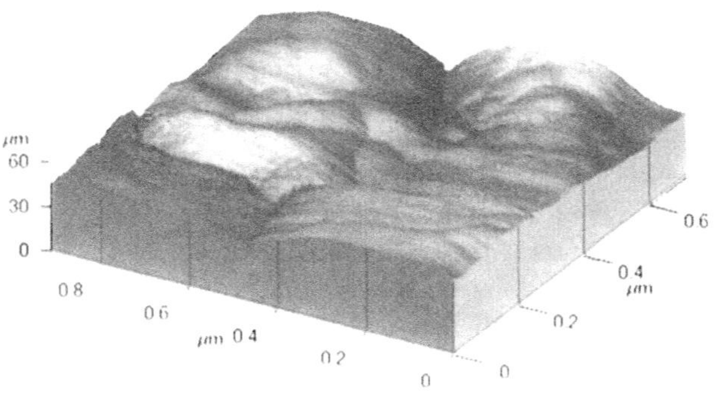

Figure 3. Morphology of uncoated ZnS powder particles.

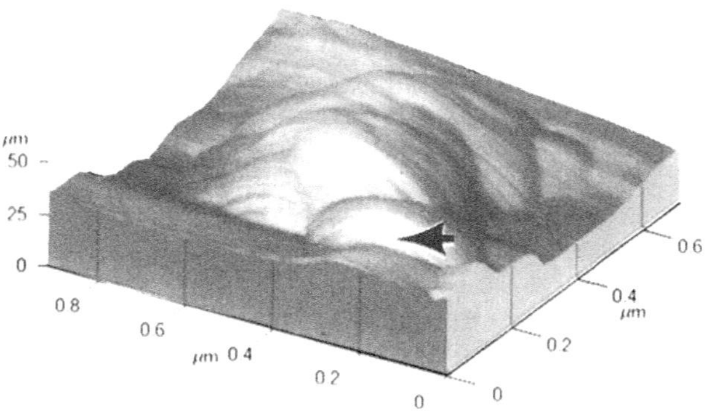

Figure 4. Morphology of coated ZnS powder particles.

diamond thin film has different characteristics. In order to evaluate objectively the surface characterization, imaging of 10 to 20 locations on the sample is needed. An average root mean square (rms) roughness was calculated based on scanning 20 locations, an area of 10-80 μm^2.

Uncoated and Coated ZnS Powder

Coated powders showed distinguished surface morphology from that of uncoated powders in the ZnS samples. The surface topography is shown in Figures 3 and 4. Most particles of the uncoated powder sample in Figure 3 revealed some dips on the surface and particles seemed to be deeply immersed into a thermoplastic adhesive. The particle size in Figure 3 is about 500 nm in diameter and dip size is about 60 nm in width. But these sizes differed from particle to particle since only a portion of the particle surface was imaged by AFM. Three dimensional surface topography showed incomplete conformal deposition layers on ZnS particles, e.g., indicated by the arrows (Figure 4). The incompletely conformal deposition layer seemed to be a convex-concave spherical coating.

Surface characteristics are closely related to the force in AFM measurements. We think that, in general, the major contribution to AFM imaging is from the van der Waals forces, ionic force and capillary force for atomically flat surface. Macroscopic friction and tension also play an important part, especially, when we evaluate very rough and hard surfaces. We observed that if powder particles were immersed only to a small depth (shallow) into thermoplastic adhesive, they were swept away by the tip of AFM, shown in Figure 5. This phenomenon provides a new method to measure the exerted force of a tip on powder particles. We measured this force to be of the order of the 10^{-6} - 10^{-7} N, which is much greater than atomic force.

ACKNOWLEDGMENTS

The authors are indebted to D. O'blans for nice discussions on this paper and Osram Sylvania Inc, Towanda, PA, for providing the powder samples.

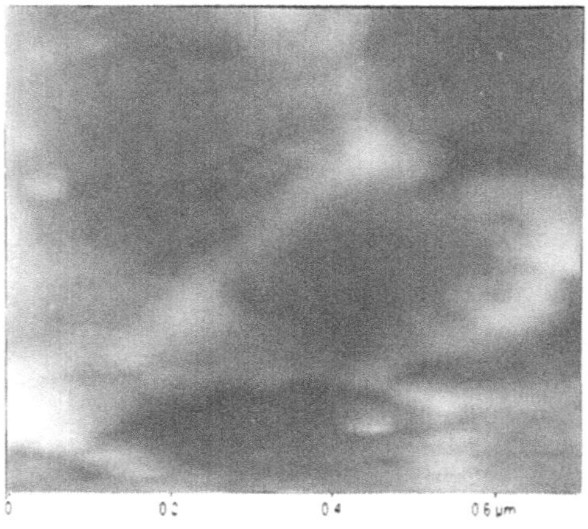

Figure 5. Craters of swept powder particles by a tip of AFM

REFERENCES

1. G. Binnig and C. F. Quate, Atomic force microscope, *Phys. Rev. Lett.*, 56:930-933 (1986).
2. F. Ohnesorge and G. Binnig, True atomic resolution by atomic force microscopy through Repulsive and attractive forces, *Science*, 260:1451-1456 (1993).
3. L. Vazquez, O. Sanchez, and J. M. Albella, Scanning tunneling microscopy morphological study of the first stages of growth of microwave chemical vapor deposited thin diamond films, *J. Vac. Sci. Technol.* 12B:1-7 (1994).
4. V. Baranauskas, M. Fukui, C. R. Rodrigues, N. Parizotto, and V. J. Traua-Airoldi, Direct observation of Chemical vapor deposited diamond films by atomic force microscopy, *Appl. Phys. Lett.*, 60:1567-1569 (1992).
5. L. F. Sutcu, C. J. Chu, M. S. Thompson, R. H. Hauge, and J. L. Margraue, Atomic force microscopy of (100), (110), and (111) homoepitaxial diamond films, *J. Appl. Phys.*, 71:5930-5940 (1992).
6. F. Lin, D. Meier, D. Hoyt, T. Van Seambrouck, and J. Leckenby, Atomic force microscopy. A tool for analysing coatings, *Material World* 1:393-395 (1993).
7. G. Persch, C. Born, H. Engelmann, K. Koehler, and B.Utesch, Application of scanning probe microscopes in technology and manufacture, *Scanning*, 15:283-290 (1993).

CURRENT VERSUS VOLTAGE CHARACTERISTICS FOR DEPOSITION AND REMOVAL OF GOLD NANOSTRUCTURES ON A GOLD SURFACE USING SCANNING TUNNELING MICROSCOPY

J.M. Perez[1] and J.L. Large[1,2]

[1]University of North Texas
Department of Physics
Denton, Texas, 76203
[2]Austin College
Department of Physics
Sherman, Texas, 75090

Abstract: We report the deposition and removal of gold nanostructures 50-200 Å in diameter on a gold surface using scanning tunneling microscopy (STM) by applying a dc voltage to the tip. The dc voltage is applied using tunneling current versus voltage (I-V) spectroscopy. We observe the deposition of a gold nanostructure at a threshold voltage of approximately 2.8 V. The I-V curves show a large current after deposition consistent with a tip-sample connection. The fact that deposition occurs using a dc voltage supports this tip-sample connection. We also observe the removal of an existing nanostructure which leaves a pit behind on the surface by applying a dc voltage with the tip directly above the nanostructure. In this case, the I-V curves show no current consistent with no tip-sample connection being formed. We conjecture that an electric-field-induced force picks up the nanostructure in whole onto the tip. We propose that occasional pit formation is the result of a tip-sample connection which breaks while the tip is under bias.

INTRODUCTION

Scanning tunneling microscopy (STM) has recently been used to deposit gold nanostructures on a gold surface by applying a short voltage pulse to the tip.[1] The physical mechanisms responsible for this phenomenon are the subject of considerable current interest. Possible mechanisms that have been proposed are field emission of atoms from the tip[1-3] and tip-sample contact.[4] In this paper, we report, for the first time, the deposition of gold nanostructures by applying a dc voltage instead of a voltage pulse to the tip. The dc voltage is applied using tunneling current versus voltage (I-V) spectroscopy.

Using this technique, the current between the tip and sample can be monitored. We observe a large current, which clips the current amplifier after deposition. We have attributed this behavior to a tip-sample connection.[5] •We also report the removal of an existing nanostructure, which leaves a pit behind on the surface by applying a dc voltage with the tip directly above the nanostructure. In this case, no current is observed indicating that no tip-sample connection is formed. We attribute the removal to an electric-field-induced force between the tip and nanostructure, which picks up the nanostructure in whole onto the tip. We propose that the occasional production of a pit instead of a nanostructure when a voltage pulse is applied, which has been reported,[1-3] is caused by a tip-sample connection that breaks while the tip is biased. Our model provides a self-consistent explanation for both nanostructure and pit formation.

Our experiments were performed in air using a commercial scanning tunneling microscope with a digital feedback loop.[6] Gold tips were prepared by electrochemically etching gold wire 0.02 inches in diameter in concentrated HCl acid using voltages of 3.0-3.5 V ac.[7] Submicrometer radii at the tips were observed using scanning electron microscopy. The gold substrate was prepared by heating the end of a gold wire 0.02 inches in diameter in a butane-nitrous oxide flame until a ball formed at the end. The ball exhibited atomically flat (111) oriented facets,[8] observed using STM. Typical sample voltages and constant tunneling currents used for imaging were 0.1-0.2 nA and 75.0-150.0 mV, respectively. I-V spectroscopy was performed at a particular sample location by disengaging the feedback loop and measuring the tunneling current as a function of tip-sample voltage at a fixed tip-sample distance. Locations on the surface at which to perform I-V spectroscopy were input to a computer. When the tip reached such a location during imaging, the computer disengaged the feedback loop, performed I-V spectroscopy, and resumed imaging. The starting and ending voltages and duration of the voltage scan were adjustable parameters.

Previous reports involving deposition of gold nanostructures on a gold surface applied a voltage pulse on the order of a few hundred nanoseconds to the tip.[1-4] The voltage pulse was capacitively coupled to the STM electronics and did not disturb the feedback loop. We were able to reproduce these results using a similar set-up. In this paper, we report the deposition of similar nanostructures by applying a slowly varying dc voltage to the tip using I-V spectroscopy. This technique has the advantage that we can measure the current during deposition.[5] Figure 1(a) shows an STM image with three locations indicated by the numbers 1,2, and 3 where I-V spectroscopy was performed, as described above. A starting voltage of -3 V, ending voltage of 3 V, and voltage scan time of 10 s were used. At these locations, we see the deposition of a circular nanostructure approximately 150 Å in diameter. Only the top halves of the nanostructures are observed in the figure because the nanostructures did not exist before the I-V spectroscopy was performed. This indicates that the deposition is highly symmetrical about the tip. Also observed at the bottom of Figure 1(a) are nanostructures deposited during previous scans to serve as references. Figure 1(b) shows an STM image of the same region in Figure 1(a) taken after the depositions. Three symmetrical nanostructures are clearly observed. Using voltage scans with different negative starting voltages and ending voltages of 0 V, we measured a negative threshold voltage of approximately -2.8 V. Starting voltages with absolute values less than 2.8 V did not result in deposition. Using voltage scans with different positive starting voltages, we measured a positive threshold voltage of approximately 2.9 V. These threshold voltages are similar to those reported in Ref. 1 for comparable tip-sample distances. Negative tip biasing was slightly more reproducible than positive tip biasing in depositing nanostructures.

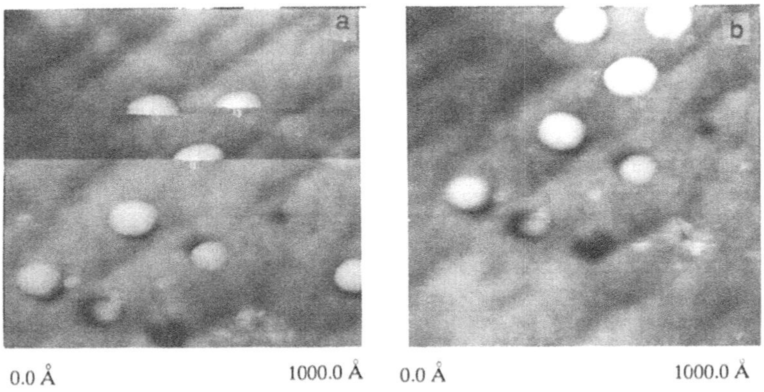

Figure 1. (a) Deposition of three nanostructures at the top of the figure using I-V spectroscopy. The numbers 1,2, and 3 denote positions of the tip during I-V spectroscopy. (b) Same area as in (a) imaged after the depositions showing three symmetrical nanostructures.

Figure 2 shows the I-V curve for the first deposition shown in Figure 1(a). The current is observed to clip the current amplifier, which clips for currents larger than 10 nA, during the entire voltage sweep. The large current continues down to 0 V and abruptly reverses polarity when the voltage reverses polarity. This can be explained by a connection between the tip and sample which has the effect of shorting the tip to the substrate, as we have previously reported.[5] This connection can explain why a dc voltage produces the same size nanostructures as a short voltage pulse. If we assume that an electric field between the tip and surface initiates the deposition, then the instant the tip makes contact with the surface this electric field is reduced to zero, even though a voltage is being applied, stopping any further deposition. Thus a dc voltage would have the same effect as a voltage pulse, as we observe. Figure 2 also shows that the connection persists for the entire 10 s duration of the voltage scan. The observed nanostructure is therefore the result of the breaking of the connection when the imaging scan is resumed.

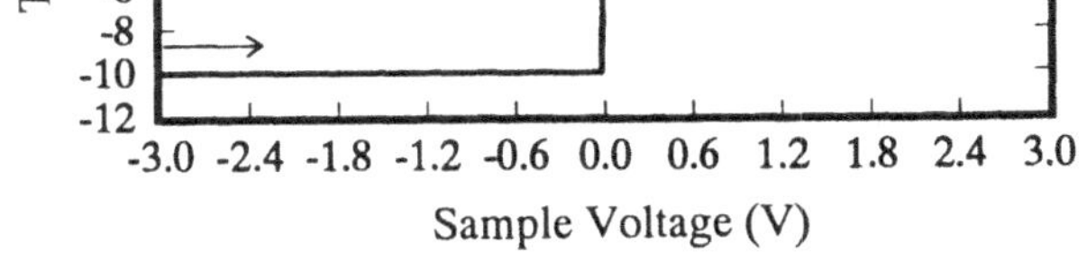

Figure 2. I-V curve for the first deposition shown in Figure. 1 (a). The starting voltage was -3 V, ending voltage 3 V, and the voltage scan time 10 s. A current larger than 10 nA which clips the current amplifier is observed down to 0 V. The clipping reverses polarity when the voltage reverses polarity. This can be explained by a connection between the tip and surface which shorts the tip to the substrate. Arrows indicate direction of the voltage scan.

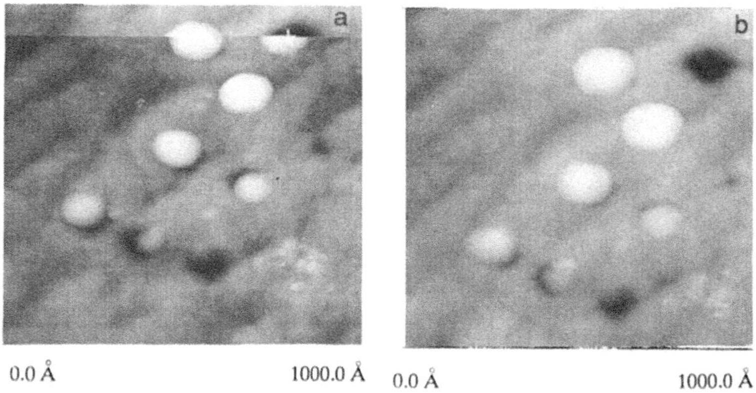

Figure 3. (a) Removal of a previously deposited nanostructure at the top right-hand side of the figure by performing I-V spectroscopy with the tip directly above the nanostructure. The number denotes the position of the tip. The nanostructure is not observed after the imaging scan is resumed. (b) Image of the same area shown in (a) after the removal showing that the nanostructure has disappeared and a pit exists where the nanostructure used to be.

Figure 3(a) shows an STM image where I-V spectroscopy is performed directly above the right-most nanostructure in Figure 1(b). A starting voltage of -3 V, ending voltage of 0 V, and scan time of 10 s were used. After the imaging scan is resumed, we observe that the nanostructure has disappeared. This is confirmed in Figure 3(b), which shows an STM image of the same region after the removal showing a pit where the nanostructure used to be. By using different starting voltages, threshold voltages for removal were measured to be -2.9 V and 3.0 V. Figure 4 shows the I-V curve for the removal shown in Figure 3(a). In contrast to the case of deposition, no current is observed during the entire scan. Thus a tip-sample connection is not formed. Since the surface surrounding the location is not altered after removal, we conjecture that the tip has picked up the nanostructure in whole. If this conjecture is true, then there is an electric-field-induced force between the tip and nanostructure. We note that such a force would also explain nanostructure deposition. In this case, the force elongates the tip until a connection with the surface is made which reduces the electric field to zero. In the case of removal, we conjecture that nanostructure or tip do not elongate to form a connection because the nanostructure is not sufficiently attached to the substrate and is instead lifted towards the tip.

Field evaporation, on the other hand, would occur between a loosely attached nanostructure and the tip and should therefore result in a connection, which is not observed. To provide evidence that the nanostructures are picked up by the tip, we deposited a nanostructure immediately after removing one. In these cases, the deposited nanostructure was significantly larger in size than the typical nanostructure deposited after a deposition. Figure 5(a) shows such a nanostructure deposited after a removal at the bottom of the figure. This nanostructure is larger than the nanostructure deposited afterwards at the top of the figure. Figures 5(b) and 6 show STM top-view and topographic images, respectively, of these nanostructures showing a volume difference of approximately a factor of two.

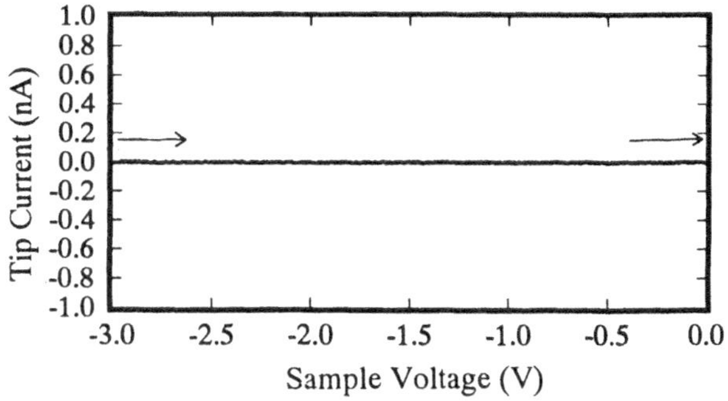

Figure 4. I-V curve for the nanostructure removal shown in Figure 3(a). The starting voltage was -3 V, ending voltage 0 V, and scan time 10 s. No measurable current is detected during the entire scan indicating the no tip-sample connection is formed. We conjecture that the nanostructure is picked up in whole onto the tip.

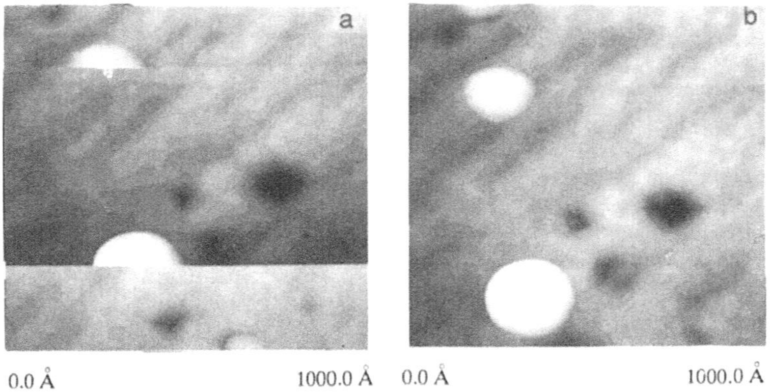

Figure 5. (a) Deposition of a nanostructure at the bottom of the figure indicated by the number 1 after removal of a nanostructure. Subsequent deposition of another nanostructure during the same imaging scan at the top of the figure indicated by the number 2. (b) Image of the same area after the depositions. The first deposited nanostructure is significantly larger than the second providing evidence that nanostructures are removed by being picked-up by the tip.

Additional evidence that nanostructures are picked up by the tip is provided by Figures 3(a) and (b). Figure 3(b) shows an image after the nanostructure removal shown in Figure 3(a). The nanostructures in Figure 3(b) are observed to be about 15% larger in diameter than in Figure 3(a). However, the distance between centers of nanostructures and between shallow surface features in both figures are the same. In addition, the pit at the bottom of Figure 3(b) is smaller in diameter than in Figure 3(a). These observations are consistent with a tip which has increased in tip diameter after the removal. We note that these effects are not observed in Figures 1(a) and (b) after depositions.

The formation of a tip-sample connection and the removal of a nanostructure by application of a voltage with the tip directly above the nanostructure provide an explanation

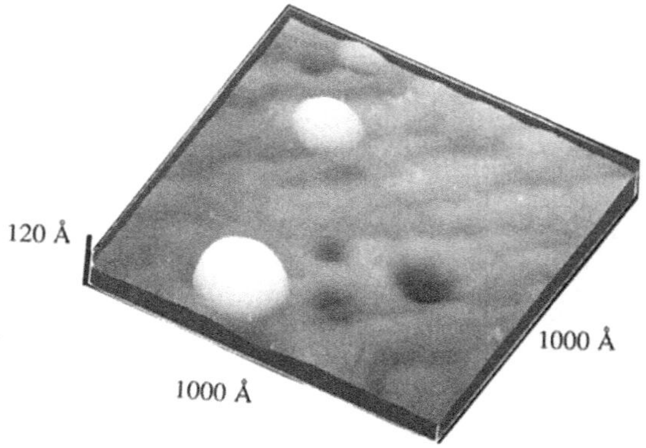

Figure 6. Topographic STM image of Figure 5(b). The bottom nanostructure is observed to have a volume approximately twice that of the top nanostructure providing evidence that the tip picks up nanostructures during removal.

for the reported[1-3] phenomenon that occasionally a pit instead of a nanostructure is formed when a voltage pulse is applied. We propose that in pit formation a tip-sample connection is first formed which then breaks while a voltage exists between the tip and surface. The nanostructure resulting from the break is then removed leaving a pit behind, as we have shown. Thus our model provides a self-consistent explanation of both nanostructure and pit formation.

Deposition using gold tips in air was attempted on other substrates including Si and GaAs. Deposition of very large structures on the order of 1500 **Å** could be achieved at voltages of about 5.0 V. Thus the small size observed for deposition on gold may be due to other tip-sample interactions such as wetting, as has been recently theoretically discussed.[9] By using tips made of metals that wet Si or GaAs it may be possible to deposit nanostructures on these substrates for device applications.

In summary, we have presented a self-consistent model for nanostructure and pit formation. Nanostructures are formed when a tip-sample connection is formed which breaks when the tip is not under bias. Pits are formed when a tip-sample connection is formed which breaks when the tip is under bias. These results provide evidence that an electric-field-induced force between the tip and sample instead of field emission is responsible for deposition of nanostructures.

ACKNOWLEDGMENTS

This work was supported in part by the National Science Foundation under Award No. DMR-9112940 and the National Aeronautics and Space Administration under Award No. NAG-1-1468. The authors acknowledge the assistance of John Burciaga for useful conversations and Dwight Maxim for assistance in building the electronic apparatus.

REFERENCES

1. H.J. Mamin, P.H. Guethner, and D. Rugar, Atomic emission from a gold scanning-tunneling-microscope tip, *Phys. Rev. Lett.*, 65:2418-2421(1990).
2. S.E. McBride and G.C. Wetsel, Jr., Nanometer-scale features produced by electric-field emission, *Appl. Phys. Lett.*, 59:3056-3058 (1991).
3. C.S. Chang, W.B. Su, and T.T. Tsong, Field evaporation between a gold tip and a gold surface in the scanning tunneling microscope configuration, *Phys. Rev. Lett.*, 72:574-577(1994).
4. J.I. Pascual, J. Mendez, J. Gomez-Herrero, A.M. Baro, N. Garcia, V. T. Binh, Quantum contact in gold nanostructures by scanning tunneling microscopy, *Phys. Rev. Lett.*, 71:1852-1855 (1993).
5. J.L. Large and J.M. Perez, Current versus voltage characteristics for deposition and removal of nanometer gold structures from a gold substrate using STM, Spring 1992 Meeting of the Texas Section of the American Physical Society (1992).
6. Burleigh Instruments, Inc. Rochester, NY.
7. A.J. Melmed, The art and science and other aspects of making sharp tips, *J. Vac. Sci. Technol.*, B9:601-608 (1991).
8. J.Schneir, R. Sonnenfeld, O. Marti, P.K. Hansma, J.E. Demuth, and R.J. Hamers, Tunneling microscopy, lithography, and surface diffusion on an easily prepared, atomically flat gold surface, *J. Appl. Phys.*, 63: 717-721 (1988).
9. U. Landman, W.D. Luedtke, N.A. Burnham, R.J. Colton, Atomistic mechanisms and dynamics of adhesion, nanoindentation, and fracture, *Science*, 40:454-461(1990).

ATOMIC FORCE MICROSCOPY OF ION-BEAM MODIFIED CARBON FIBERS

Pearl W. Yip and Sin-Shong Lin
U.S. Army Materials Research Laboratory
Watertown, MA 02172-0001

Abstract: Several commercial PAN and pitch based carbon fibers were modified by a low voltage ion source of nitrogen or oxygen in the range of 400-1000 eV. The fiber reinforced epoxy composites made from the treated fibers were found to have higher interfacial strength than those made from the untreated ones. The improvement of adhesive strengths obtained from the transverse tensile stress measurement reached as high as three to four folds of the original fibers. This result could be explained from increased roughness and enhanced surface functionalities. The scanning electron microscope, which traditionally is used for surface characterization, fails to detect the mild surface alteration caused by the bombardment of the low-energy ion beams. Thus, the atomic force microscope (AFM) with near-atomic resolution might provide us the information needed for the understanding of ion-modified morphology. In the present investigation, the AFM instrument, manufactured by TopoMetrix is used in the characterization of the modified fibers. Preliminary AFM results indicate significant morphology and topography changes on the treated surfaces.

INTRODUCTION

Carbon fibers, due to their special properties, such as high specific strength and modulus, have been used in the fabrication of many structural composites. However, the mechanical properties of these composites are determined largely by the adhesive bonding between the fiber and the matrix material. To improve the strength of a particular composite, one needs to understand better the interface of the reinforcing fiber and the matrix material.

The U.S. Army Research Laboratory (ARL)—Watertown has tried to modify the surfaces of various carbon fibers through many means to improve the activity of the fibers.[1,2] In this paper, four commercial carbon fibers of different origins are surface treated by nitrogen and oxygen ion beams to enhance the interface bonding through increased surface chemical activity and morphology. X-ray photoelectron spectroscopy (XPS) is used to detect a wide variety of nitrogen and oxygen functionalities that exist on the fiber surfaces produced by the ion beam treatments. Transverse tensile stress (TTS) of the fiber reinforced epoxy composites is used for strength evaluation.

Atomic Force Microscopy/Scanning Tunneling Microscopy 2
Edited by S.H. Cohen and M.L. Lightbody, Plenum Press, New York, 1997

Three different adhesive mechanisms are known to exist at the interface[3]: mechanical interlocking, chemical binding and weak attractive forces, e.g., dipole-dipole, van der Waals, and electrostatic interactions. In our earlier study, the acid oxidized carbon fiber after hydrogen reduction yielded approximately the same magnitudes of TTS as other treated fibers. This observation suggested that the oxidative treatment produced not only more active sites on the surface but also a roughened surface by etching and therefore enhanced simultaneously mechanical interlocking.[1]

Scanning electron microscopy was used to characterize the treated surfaces but was found insufficient to detect the small morphology changes caused by the mild surface oxidative treatments. Since the first commercial atomic force microscope (AFM)[4] became available in 1990, several scientists [5-8] have utilized the technology and shown the potential of AFM for studying the minor morphology changes of plasma-treated carbon and polymer fibers. For the present study, we also attempt to use AFM to elucidate the mechanisms that govern the interfacial adhesion of our ion-treated carbon fiber reinforced epoxy composites.

EXPERIMENTAL

Two PAN-based fibers, represented by Hercules IM6 (HCF) and Toho IM600 (TCF), and two pitch-based fibers, represented by Osaka Donacarbo-500 (OF5) series and Tonen HM (PCF) were used in experiment. The fiber specimens were prepared by cutting fiber strands 4" long followed by clamping the ends of strands on the perimeter of two wheels, which were fixed to a slowly rotating shaft normal to the ion. In the ion beam processing, nitrogen or oxygen gas was first ionized and then accelerated toward the fiber target at energies less than 1 kV, and a current density approximately 1 mA/cm^2.

The composite test specimens were prepared with the treated fibers and epoxy matrix. The epoxy resin consisted of 100 parts EPON 828 resin and 40 parts EPON V-40 curing agent, supplied by Miller Stephenson Chemical Co. The fiber tows were impregnated with epoxy under pressure followed by curing in an oven at 100° C overnight. This composite sheet was made into the TTS test specimen by slicing across the fiber direction into strips of approximately 36-mil width with a diamond wafer blade. Transverse tensile stress was measured with the Tinius Olson Model 1000 tensile tester using a one-pound load cell.

The fiber surfaces were examined by X-ray photoelectron spectroscopy (XPS)[10] using either PHI Model 548 ESCA/Auger instrument made by Physical Electronic Ind., Inc., or XSAM800 instrument made by Kratos Analytical Co. Two surface properties were obtained: (a) Atomic surface concentration: calculated from the integrated area under the N1s, O1s and C1s peaks in XPS spectra, and (b) abundance of functional groups obtained from the peak profile syntheses.

AFM images were collected by using a TopoMetrix TMX2000 Explorer Atomic Force Microscope at ambient environment. Scanners with a scanning range of 12 or 2 micrometers were used. Fiber specimens were mounted on sample disks with a thermoplastic adhesive from SPI Supplies.

RESULTS AND DISCUSSION

From the XPS analysis, we found that the introduction of oxygen and nitrogen functionalities on the fiber surface can be easily accomplished by the ion beam process. For the oxygen ion treated fibers, the major surface functional groups were the hydroxyl and ether type linkages (90%), and carbonyl group and acid as the minorities (10%). In addition to the carbon and oxygen, trace elements, such as nitrogen and silicon, are also detected in the PAN-based fibers. Nitrogen is an intrinsic component of the fiber precursor, while silicon is probably an impurity derived from silicon pump oil. For the nitrogen ion treated fibers, functional groups such as amide/imide, amino/nitrile and nitroso groups were found. The formation of -OCN- (amide/imide) group is more favorable than the amino/nitrile groups at higher ion energies or longer radiation time.

The adhesive strengths (TTS) of the ion treated fibers were found to be two to four times compared to the untreated fibers (Figure 1). The strength improvement is found to be more pronounced in the highly graphitized pitch-based fibers. Longer exposure time

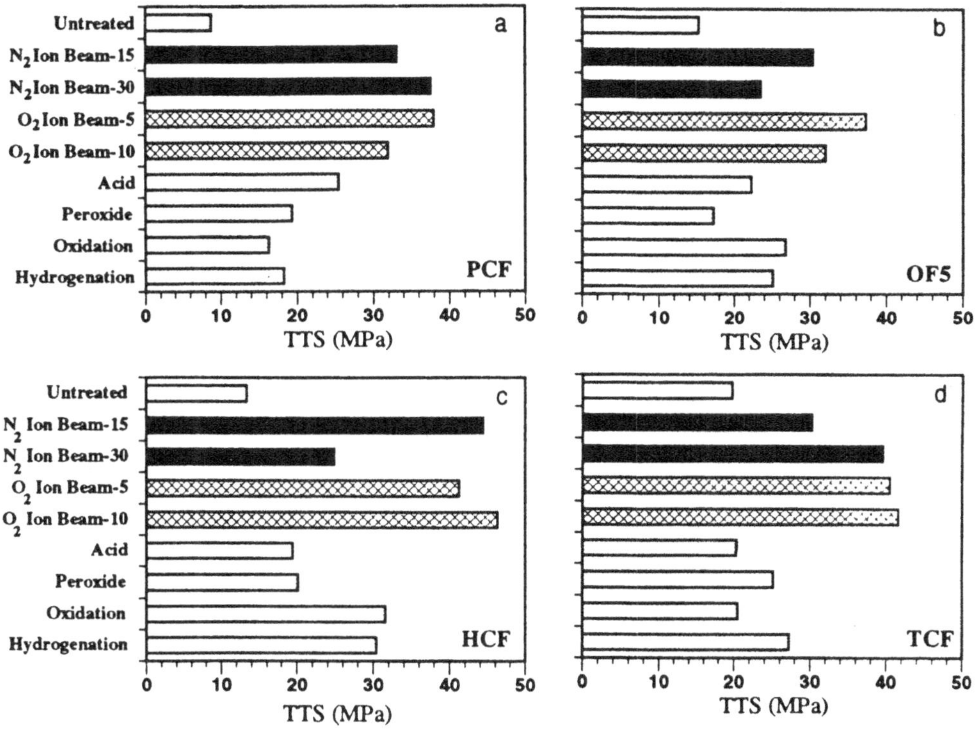

Figure 1. Transverse tensile strengths (TTS) of carbon fibers before and after various types of surface treatments for (a) PCF (b) 0F5 (c)HCF and (d) TCF fibers.

(>60 minutes) was found to deteriorate the fibers extensively. Optimum treatment time is needed to be determined for each individual fiber. From TTS comparisons, nitrogen functional groups are less effective than oxygen groups in enhancing the interfacial bonding. The abundant surface functional groups produced by the ion beam treatment as well as the increased surface roughness from etching are believed to be the major contributing factors for the higher TTS results. More detailed discussion of XPS and TTS results can be found in reference 2.

The AFM images of untreated and the treated PAN- and pitch-based fibers were collected recently (Figures 2(a) to (c) and 3(a) to (c)). At low magnification, the extrusion ridges and grooves can be observed on both types of fibers. The treated fibers appear to have rougher surfaces than the untreated ones. In our preliminary quantitative roughness analyses of the surface profiles, the average roughness (Ra) and peak-to-valley measurements for one-micrometer scanned areas are recorded and shown in Table 1. Visually, both types (PAN and pitch) of carbon fibers appear to be significantly altered after the ion beam treatments. The nitrogen ions are observed to affect the fiber surface differently compared to the oxygen ions.

The nitrogen-ion-treated samples show the most enhancement in roughness, and features on the treated surfaces also appear to be more granular or grainlike. The fuzzy or soft materials on the untreated fiber surfaces may have been knocked off by the energetic ions. Notable pitting formation is observed on the oxygen-treated PAN fibers. Although the averaged roughness values (Ra) of the treated fibers are sometimes similar to the untreated ones, the peak-to-valley values from the surface profiles are higher for fibers after ion treatments. Since for this experiment, only the standard Si_3N_4 tips having a tip curvature of 20-40 nm were used, pittings that were smaller than 40 nm might not be measured.

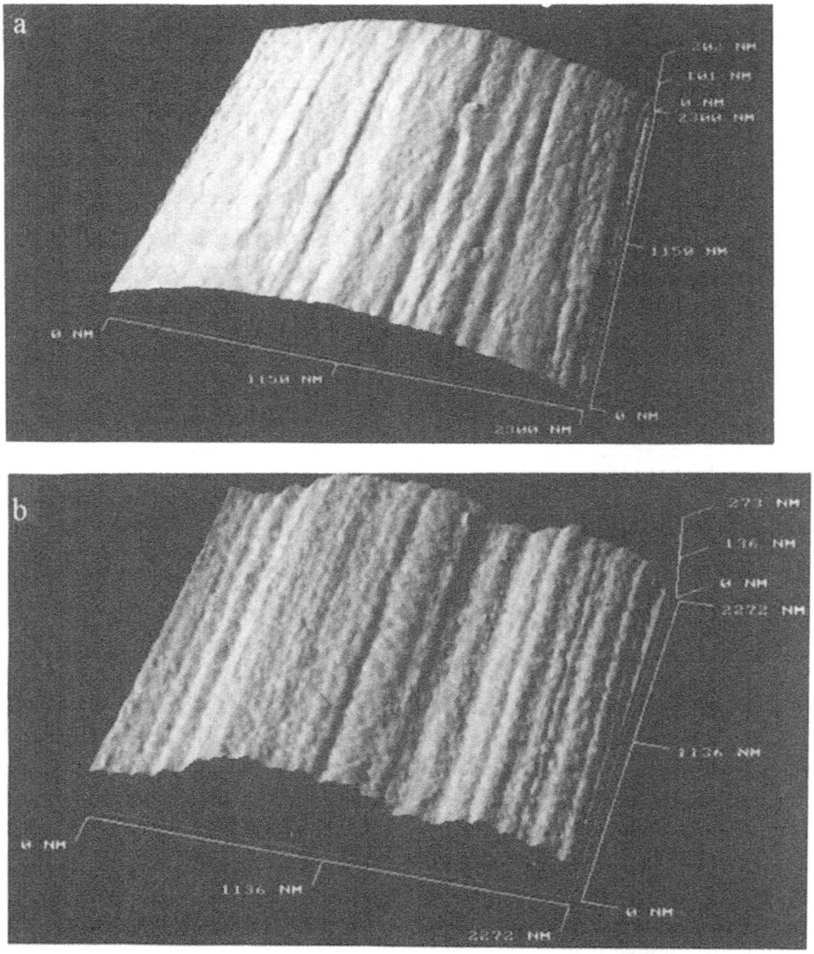

Figure 2. AFM images of PAN-based carbon fibers of (a) untreated fibers and (b) the nitrogen-ion-treated fibers.

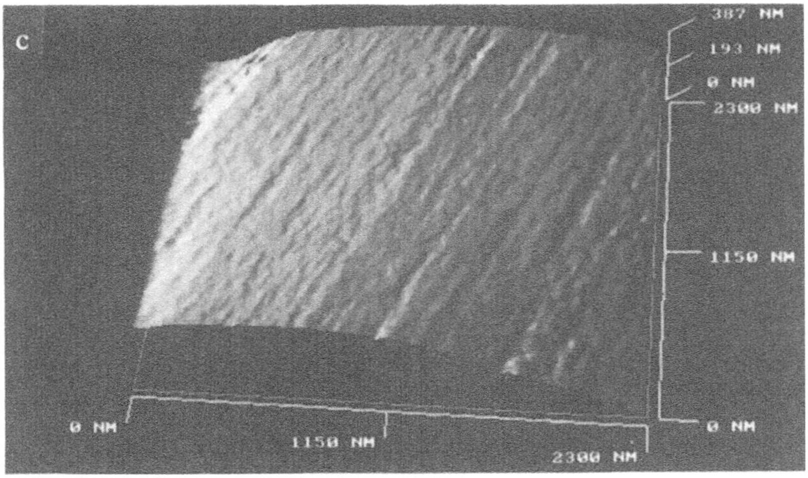

Figure 2. (c) AFM image of the oxygen-ion-treated PAN-based carbon fiber.

Table 1. Preliminary Roughness Values of Carbon Fibers Before and After Ion Treatments

Sample	Precursors	Treatment	Roughness (Ra)	p-v Heights
HCF	PAN	None	11 nm	4 - 18 nm
HCFNA	PAN	Nitrogen ion	58 nm	20 - 50 nm
HCFOF	PAN	Oxygen ion	13 nm	10 - 45 nm
PCF	Pitch	None	10 nm	8 - 36 nm
PCFNA	Pitch	Nitrogen ion	8.2 nm	7 - 42 nm
PCFOF	Pitch	Oxygen ion	8 nm	6 - 36 nm

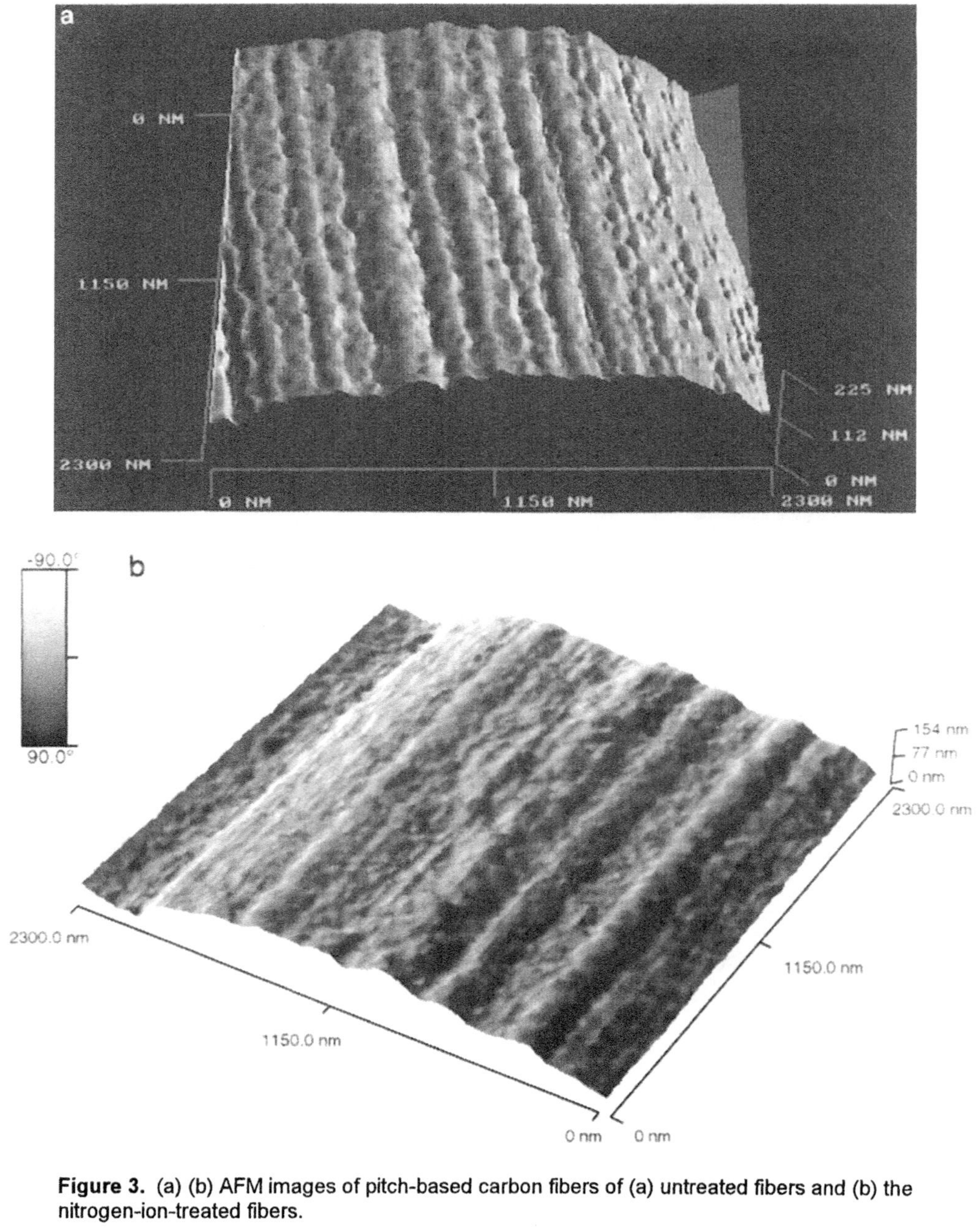

Figure 3. (a) (b) AFM images of pitch-based carbon fibers of (a) untreated fibers and (b) the nitrogen-ion-treated fibers.

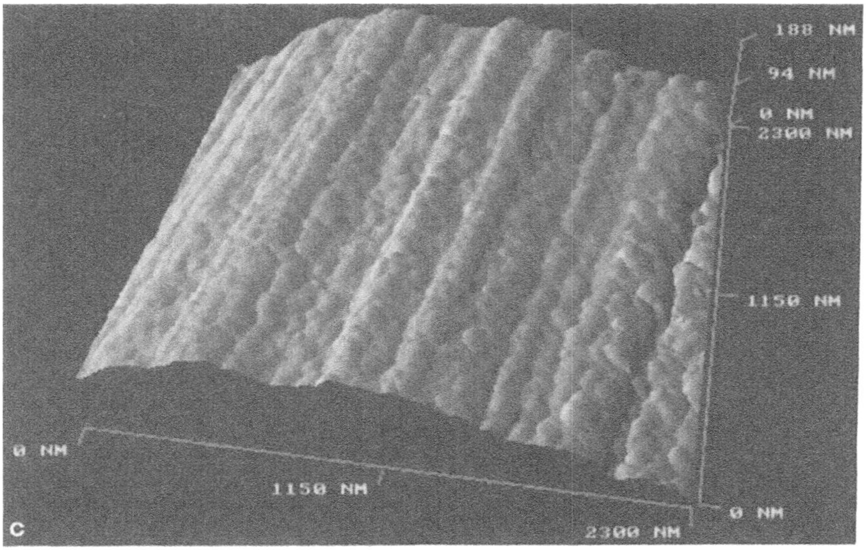

Figure 3. (c)AFM images of an oxygen-ion-treated pitch-based carbon fiber.

ACKNOWLEDGMENTS

We wish to thank Ray Eby and George Collins of TopoMetrix for the most valuable discussions on imaging carbon fibers and assistance in obtaining additional AFM photos.

REFERENCES

1. P. W. Yip and S. S. Lin, "Effect of Surface Oxygen on Adhesion of Carbon Fiber Reinforced Composites," *in:* Material Research Society Symposium Proceedings, 170: 339-344 (1990).
2. S. S. Lin and P. W. Yip, "Improved Mechanical Strengths of Epoxy Composites Obtained From Ion Beam Treated Carbon Fibers," *in* Material Research Society Symposium Proceedings, 318: 381-386 (1994).
3. E. Fitzer, "Carbon Fibers and Their Composites," Springer-Verlag, Berlin, NY Tokyo (1983).
4. G. Binnig, C.F. Quate, and C. Gerber, Atomic force microscope, *Phys. Rev. Lett.*, 56: 930-933 (1986).
5. W. P. Hoffman, Scanning probe microscopy of carbon fiber surfaces, *Carbon*, 30: 315-331 (1992).
6. S. N. Magonov, A. Ya. Gorenberg, and H. J. Cantow, Atomic force microscopy on polymers and polymer related compounds, *Polymer Bulletin*, 28: 577-584 (1992).
7. R.J. Smiley and W.N. Delgass, AFM, SEM and XPS Characterization of PAN-based carbon fibres etched in oxygen plasma, *J. Mater. Sci.* 28: 3601-3611 (1993)
8. Ittipol Jangchud, A.M. Serrano, R.K. Eby, P.E. Klunzinger, and M.A. Meador, "Investigation of Carbon Fiber Surfaces/Interfacial Bonding", *in:* 21 Biennial Conference on Carbon, 197-199 (1993).
9. S.S. Lin and P. W. Yip, "Surface and Adhesion Studies of Carbon Fiber Modified by Low Energy Ion Beam", *in:* Proc. of IUMRS-ICAM-93 Composite, Tokyo Japan, September 1993.

INDEX

The numbers following each entry refer to the first page of individual papers.

GPSR Compliance

The European Union's (EU) General Product Safety Regulation (GPSR) is a set of rules that requires consumer products to be safe and our obligations to ensure this.

If you have any concerns about our products, you can contact us on ProductSafety@springernature.com

In case Publisher is established outside the EU, the EU authorized representative is:

Springer Nature Customer Service Center GmbH
Europaplatz 3
69115 Heidelberg, Germany

Batch number: 10755809

Printed by Printforce, the Netherlands